Frederick George Kenyon

The Palaeography of Greek Papyri

Frederick George Kenyon

The Palaeography of Greek Papyri

ISBN/EAN: 9783337086473

Printed in Europe, USA, Canada, Australia, Japan

Cover: Foto ©berggeist007 / pixelio.de

More available books at **www.hansebooks.com**

THE

PALAEOGRAPHY OF GREEK PAPYRI

KENYON

HENRY FROWDE, M.A.

PUBLISHER TO THE UNIVERSITY OF OXFORD

LONDON, EDINBURGH, AND NEW YORK

THE

PALAEOGRAPHY

OF

GREEK PAPYRI

BY

FREDERIC G. KENYON, M.A.

LATE FELLOW OF MAGDALEN COLLEGE .
HON. PH.D. (HALLE), HON. D.LITT. (DURHAM)
ASSISTANT KEEPER OF MANUSCRIPTS, BRITISH MUSEUM

WITH TWENTY FACSIMILES
AND A TABLE OF ALPHABETS

Oxford
AT THE CLARENDON PRESS
1899

Oxford
PRINTED AT THE CLARENDON PRESS
BY HORACE HART, M.A.
PRINTER TO THE UNIVERSITY

PREFACE

THE substance of the following treatise was originally
written as a dissertation for the Conington Prize in 1897,
and I had at first intended to defer its publication for some
time, in view of the accessions of new material which the
excavations conducted each year in Egypt are continually
bringing to light. It was represented, however (and that
by one whose judgement I was bound to respect), that,
although the work might gain in absolute accuracy by
such a delay, yet its usefulness to students would be greater
if it were published now, as enabling them to assimilate
the new material for themselves. Consequently it is without any idea of having reached finality, but rather as an
assistance to investigation, that this book is now offered to
the consideration of students of palaeography.

It is, in fact, an essay in the strict sense of the term—an
attempt to formularize and classify the results of a number
of discoveries, most of which have occurred quite recently.
Hence, it will be understood, the conclusions stated in it are
in many instances only the impressions of a single student
of the subject, and do not (as is the case with treatises on
the palaeography of vellum manuscripts) express the con-
sensus of opinions of many experts and many ages. The
whole subject is new; fresh materials are coming to light

year by year, and much of that which is already extant
has not been published in such a form as to make it available for students at a distance from the originals. Hence,
although the wealth of the British Museum in papyri (and
especially in literary papyri) gives a considerable advantage
to a student whose work lies in that sphere, it is possible
that the experts of Berlin and Vienna may sometimes have
been led by their experience among the yet un-photographed
documents in those collections to conclusions different from
those which are expressed in the following pages. Still,
the amount of accessible material is now so great, and
spread over such wide periods of time, that inductions may
be drawn from it with a fair amount of certainty; and
with regard to the literary papyri, which form by far the
most important branch of the subject, fortune has hitherto
brought nearly every manuscript of the first importance
to the British Museum; so that it seems justifiable to try
to state some general results and principles to which
a study of these materials seems to lead us, in the hope
that the development of this branch of palaeographical
science may thereby be facilitated.

In conclusion I have to thank the Delegates of the
University Press for undertaking the publication of this
volume, and especially Prof. Bywater for his supervision
of it while passing through the press; and I wish also
gratefully to acknowledge the assistance of Sir E. Maunde
Thompson, K.C.B., Director of the British Museum, who
has found time to read the proof-sheets and to make many
valuable criticisms and suggestions.

F. G. K.

October 22, 1898.

CONTENTS

PLATES

— ·· —

CHAPTER I

THE RANGE OF THE SUBJECT

THE science of palaeography in its application to Greek writing upon papyrus is a development of quite recent years. Sixty years ago, Greek palaeography practically began with the fifth century of our era, and dealt almost wholly with writing upon vellum and paper. Even the two great codices which now stand at the head of the list of Greek vellum MSS. were not at that time within the scope of the palaeographer; for the Codex Vaticanus was jealously secluded in the Vatican Library, and the Codex Sinaiticus had not yet been discovered. It is true that some examples of Greek writing upon papyrus had been known since the latter part of the eighteenth century : but these, with the exception of the charred (and only very partially accessible) rolls from Herculaneum, did not contain literary works, and were merely isolated documents, or groups of documents, unconnected with one another or with the known palaeography of vellum MSS. Shortly before the middle of the present century, however, papyri began to be found in Egypt in somewhat more considerable numbers, and among them were several which contained works of

Previous position of the subject.

B

literature. The discovery of a long roll, containing three
of the lost orations of Hyperides, besides being nearly
the first example of a literary MS. on papyrus in good
preservation, was the first-fruits of the harvest of lost
authors which Egypt has yielded, and is still likely to
yield, to the explorer. Some considerable portions of
Homer were also added to the stock of available material
at about the same time. But the progress of discovery
was slow at first, and these earlier successes were not
followed for several years by much that was important
either in quantity or in quality. It is only within the
last twenty years that the stream has begun to flow
with much fullness; and only within the last seven years,
or less, has it become possible to give anything like a
continuous record of the character and development of
Greek writing during the period when papyrus was the
material mainly in use for its reception.

The recent discoveries of papyri have, in fact, added
a province of nearly a thousand years to the domain
of the palaeographer, beginning about three hundred years
before Christ, and coming down to a limit which is best
fixed by the Arab conquest of Egypt in A.D. 640. It is
only during the last three hundred years of this period
that it overlaps the sphere occupied by vellum MSS. For
six hundred years the papyri held the field alone, and only
now are we beginning to be able to realize their character
and understand the history of their development. That
history has never yet been written with any fullness of
detail; even the outlines of it have hardly been laid down.
When Gardthausen wrote his *Griechische Paläographie*, in
1879, his treatment of papyrus-palaeography was based
upon a single group of documents belonging to the second
century B.C., and two or three isolated and non-literary
pieces scattered over the centuries between that date
and the rise of vellum MSS. The great Hyperides MS.
was the principal example of a literary hand accessible to

him, and of that the age was quite doubtful. His treatment of the subject is, consequently, now out of date; and
no palaeographical treatise on the same scale has been
written to take its place. The only work in which the
new material has been dealt with is Sir E. Maunde
Thompson's *Handbook of Greek and Latin Palaeography*
(1893), and that necessarily treats of it very briefly [1].
Moreover, even since the date of that book the mass
of available material has been largely increased, and some
serious gaps in our knowledge have been filled up. Under
these circumstances, an attempt to set the material in
order, and to formulate, even if it be but provisionally,
the laws which regulated the development of Greek
writing upon papyrus, seems to be justifiable. Though
our knowledge is certainly destined to increase, and that
soon, it may be useful to mark the limits which have
hitherto been gained, and thereby to pave the way for
future progress.

The first discovery of Greek papyri in modern days was
made at Herculaneum in 1752. Before that date the only
papyri known to the palaeographer were a few survivals
from mediaeval times. Letters, papal bulls, and municipal
archives written upon papyrus in Latin existed in considerable numbers, chiefly in Italy; but in Greek little
was extant at all, and that little was in extremely bad
condition [2]. The excavations at Herculaneum, however,
brought to light a mass of charred papyrus rolls, to which
must be assigned the honour of inaugurating the new era,
though it was forty years before these began to be made

*The ear-
liest dis-
coveries of
papyri.*

[1] A still more brief survey of the subject is given by Prof. Blass in
Müller's *Handbuch der klassischen Altertumswissenschaft*, vol. i. (1891). The
last edition of Wattenbach's *Anleitung zur griechischen Paläographie* (1895
attempts little more than a bibliography of the subject.

[2] Montfaucon, writing in 1708, says 'Graecum autem papyreum codicem
nullum hactenus vidimus' (*Palaeographia Graeca*, p. 15), though he subsequently describes some much mutilated fragments in the library of
St. Martin at Tours, and refers to the existence of a few others at Vienna.

available to scholars, the first volume of facsimile texts appearing at Naples in 1793. Even then their publication proceeded very slowly, especially during the troubled times of the next thirty years; the Naples volumes were not very easily accessible to scholars in general; and the charred condition of the rolls detracted greatly from their value. Meanwhile a far more important mine was opened in 1778 by the discovery in Egypt, probably in the province of the Fayyum [1], of some forty or fifty rolls of papyrus. They were offered, by the natives who found them, to a dealer, who bought one out of curiosity, but refused the rest; whereupon they were burnt by the natives for the sake of the smell which they gave forth. The one survivor was sent to Cardinal Stefano Borgia, and placed in his Museum at Velletri, where it was edited by N. Schow in 1788 [2]. It contains nothing but a list of peasants employed in forced labour on the embankments regulating the Nile floods, and is of little importance save as the forerunner of much more valuable discoveries. For some time, however, discoveries came but slowly, and the publications of the next five-and-thirty years are few and unimportant. At last, about 1820, a new start was made. A large group of papyri was found (again, it is said, in an earthen pot) on the site of the Serapeum at Memphis; and these, which are now divided between the museums of Paris, London, Leyden, Rome, and Dresden, form the foundation of our knowledge of the palaeography of the second century B.C. Simultaneously papyri began

---

[1] The find was reported to have been made at Gizeh, in a buried earthenware pot: but since the surviving document relates to the local affairs of the village of Ptolemaidis Hormus, in the Fayyum, it is much more likely to have been found in that neighbourhood. The statements of native discoverers as to the *provenance* of papyri are not valuable as evidence.

[2] *Charta papyracea Graece scripta Musei Borgiani Velitris* (Rome, 1788), with specimen facsimiles. The document is of the end of the second century, being dated (though Schow overlooked this) in the latter part of the thirty-first year (of Commodus = A.D. 191).

to appear from other sources, and the study of the materials thus provided soon led to the publication of several volumes of considerable note in the bibliography of papyri. In 1821 Mr. W. J. Bankes acquired a papyrus roll containing the last book of the *Iliad* [1]—the first literary papyrus to be discovered, with the exception of those from Herculaneum. In 1824 the publication of the Herculaneum papyri was resumed at Oxford. Two years later came the publication of the Turin papyri by Amadeo Peyron; while in 1839 those of the British Museum were edited by Forshall, and in 1843 those of Leyden by Leemans. These three volumes contained nothing but non-literary documents: but in the years just before and after the middle of the century several important literary papyri came to light. In 1847 Mr. Arden and Mr. Harris independently obtained portions of a great roll containing three speeches of Hyperides [2]—the first previously unknown classical author to be recovered on papyrus: in 1849 and 1850 Mr. Harris obtained two portions of a MS. of the seventeenth book of the *Iliad*, and another, in book form, of books ii–iv [3]: in 1855 M. Mariette secured for the Louvre a fragment of Alcman [4]; in 1856 Mr. Stobart acquired the *Funeral Oration* of Hyperides [5]. Some small portions of Homer were also acquired for the Louvre about the same time [6].

With these discoveries the first age of papyrus-revelation may be said to have ended. A pause followed for twenty years, and then a new era was inaugurated, an era of discoveries on a large scale, in which we may be said to be still living. In 1877 an enormous mass of papyri, mainly non-literary, was unearthed about the site of

Discoveries of the last twenty years.

---

[1] Commonly referred to as the Bankes Homer; now Pap. cxiv in the British Museum.

[2] Now Brit. Mus. Papp. cviii and cxv.

[3] Now Brit. Mus. Papp. cvii and cxxvi. The former (being the earlier to be made known) is the MS. generally known as the Harris Homer.

[4] Louvre Pap. 70.   [5] Brit. Mus. Pap. xcviii.

[6] Louvre Pap. 3, 3bis, 3ter.

Arsinoë in the Fayyum, of which the greater part went
to the collection of the Archduke Rainer at Vienna, though
Paris, London, Oxford, and Berlin also had a share.

The papyri of this first Fayyum find were almost entirely
of the Byzantine period, and for the most part were very
fragmentary. Since that date, though papyri have also
come from Eshmunên (Hermopolis), the neighbourhood of
Thebes, and elsewhere, the mounds of the Fayyum have
been the most prolific source of papyri, nearly all, however,
up to the present, having been of a non-literary character.
About 1892 another very large find was made in the same
district, principally, it would appear, from the site of
a village called Socnopaei Nesus; and this time the lion's
share fell to Berlin, although the British Museum secured
a good representative selection, and others went to Vienna
and Geneva. This second great find differed from the first
in containing far more perfect documents, and in belonging
mainly to the Roman period, from the beginning of the
first to the middle of the third century after Christ.
Meanwhile the range of our knowledge had been extended
backwards by Mr. Flinders Petrie's discovery (in 1889-90)
of a number of mummy-coffins, the cartonnages of which
were composed of fragments of papyri written in the third
century B. C. Most of these were business documents; but
the literary specimens included two very valuable items
in the fragments of Plato's *Phaedo* and the lost *Antiope*
of Euripides. About the same time the British Museum
acquired a most remarkable series of literary papyri, including the lost Ἀθηναίων Πολιτεία of Aristotle, the *Mimes*
of Herodas, part of another oration of Hyperides, and
a long medical treatise, to say nothing of MSS. of
Homer, Demosthenes, and Isocrates. To these the Louvre
added in 1892 the greater part of Hyperides' masterpiece,
the *Oration against Athenogenes*. In the same year a gap
in the chronological sequence of dated papyri was filled
by the acquisition on the part of the British Museum

and the Geneva Library of a group of documents from the middle of the fourth century (the correspondence of a Roman officer named Abinnaeus); while the papyri obtained by Messrs. Grenfell and Hogarth in the years 1894–1896 have provided ample material for the palaeography of the latter half of the second century B. C., and have even extended our knowledge for some distance into the previously blank first century B. C. The winter of 1896–7, however, threw into the shade all previous discoveries, with the possible exception of those of 1890–1. The British Museum acquired a papyrus containing considerable portions of the odes of Bacchylides; M. Nicole, of Geneva, secured some fragments, small but interesting, of a comedy by Menander; while Messrs. Grenfell and Hunt, excavating at Behnesa (Oxyrhynchus) on behalf of the Egypt Exploration Fund, unearthed masses of papyri which rival in quantity, and far surpass in quality, even the great finds of Arsinoë and Socnopaei Nesus. The thousands of papyri discovered on this occasion include, along with masses of fragments, large numbers of non-literary rolls in excellent condition; and, more valuable but more tantalizing, scores of fragments of literary works, known and unknown. The full list of these discoveries has not yet been made public, but it includes the now famous 'Logia' of our Lord, and fragments of St. Matthew, Sappho (probably), Thucydides, Sophocles, Plato, and other authors, besides the inevitable Homer. It may safely be predicted that the harvest of that season will not be surpassed for many a year to come.

The general result, from the palaeographical point of view, of this series of discoveries, and especially of those of the last six years, is that we now have an almost uninterrupted series of dated documents from B.C. 270 to A. D. 680. The third and second centuries B. C. are now so fully represented that there seems little room for serious error in dealing with MSS. of these periods. For

Chronological distribution of the material.

the first century, especially the transition from Ptolemaic
to Roman writing, material was until two years ago almost
wholly wanting, and still is but scanty. From the beginning
of the Christian era the supply begins to grow plentiful
again, and from about the middle of the first century until
the second quarter of the third century dated papyri exist
in such numbers that hardly a year is without many of
them. The last half of the third century is somewhat
thinly represented; but the age of Diocletian (a turning-
point in the history of palaeography) is covered by some
recent acquisitions of Mr. Grenfell's (now in the British
Museum), and the middle of the fourth century is fully
known from the papyri of Abinnaeus at London and
Geneva, to which reference has been made above. A few
documents dated about the year 400 have lately come to
light, but the fifth century is now the darkest period in
the palaeography of papyri. The sixth century and the
first half of the seventh, on the other hand, have yielded
a huge harvest; but the proportion of precisely dated
documents is small, and the accurate apportionment of
the various types of hands cannot be made with absolute
certainty. With the Arab conquest of Egypt the practice
of Greek writing on papyrus received its death-blow, and
during the latter half of the seventh century it slowly
flickered out. No dated document of the eighth century is
in existence, except two tiny receipts in Arabic and Greek,
though one long papyrus is assigned with apparently good
reason to this period.

Inferences
as to early
use of
writing.
The end of the papyrus period is, therefore, fixed and
known; but with the beginning it is different. Recognizing
that the material which we have from the early part of the
third century B.C. is sufficiently plentiful to allow us to
carry our inferences back for a generation or two, we may
fairly say that we know how men wrote in the days of
Aristotle and Menander; but we have not yet got back
to Pindar and Aeschylus, much less to Homer or (if a less

contentious name be preferred) Hesiod. There are, however, certain broad generalizations and inferences which have a bearing even upon these distant periods. In the first place it is clear that, at the point where our knowledge now begins, writing was a well established art, practised not only by literary men and professional scribes, but by soldiers, farmers, and working men and women of all sorts. It was used for the daily affairs of life, to all appearance, not less than it was in England a century ago. It could be written with ease and elegance; it could also be written with a speed and fluency, which, if they sometimes offer serious difficulties to the decipherer, prove at least that the writer handled a ready pen. From this it is clear that it was no newly acquired art, but had already a long history behind it; how long, we have at present no sufficient evidence to say, but there is no sort of reason to be chary of generations. Further, it is obvious that the writing on papyrus bears no recognizable relation to contemporary inscriptions upon stone; and therefore care is necessary in using epigraphic evidence to determine the style of writing in the preceding centuries. The characters used in inscriptions may resemble the formal writing (or printing) of the same age; but they have no more bearing on the running hands in common use than modern tombstones have on the handwriting of to-day. Men may have carved formally and with difficulty upon stone; it does not follow that they did not write upon papyrus with ease and fluency.

So far we have made no formal distinction between literary and non-literary papyri; yet the distinction is fundamental for the study of papyrus-palaeography. It is a distinction that runs through the whole period of which we are speaking. Papyri which were meant to be books were written in quite different hands from the papyri which were meant to be documents, whether official or private. The difference is, in many cases, as marked as that between writing and print at the present day: Distinction between literary and non-literary writings.

and the development must be followed separately along
each line. A parallel may be found in the distinction
between the book-hands and the charter-hands of the
Middle Ages. The charters of the twelfth century may
bear some likeness, recognizable by the trained eye, to the
books of the same period; but from the thirteenth to
the fifteenth centuries the relationship is practically indis-
tinguishable, and a person who can date a charter of
Edward III or Henry VII with certainty may be quite at
sea with a chronicle or Bible of the same age. So with
papyri, an acquaintance with the succession of non-literary
hands only goes a little way towards enabling one to
fix the date of a literary MS. And there is this further
complication to be borne in mind, that whereas the
charter-hand of the Middle Ages is the hand of a trained
scribe just as much as the book-hand, the non-literary hand
of the papyri includes the writings of private individuals,
often very imperfectly acquainted with the use of the
pen, as well as those of official clerks. We have, in fact,
during the papyrus period what we have hardly at all
in the vellum period of palaeography, the casual every-day
writings of the common people; and consequently the
lines of classification which serve for vellum MSS. do not
apply. For this reason the palaeography of papyri forms
a branch apart, the principles of which must be stated
independently.

Greek writing upon vellum can be classified in two
broad and well-defined divisions, as uncial or minuscule,
the former being the earlier, the latter the later style,—
of course with some period of overlapping. Uncial writing
is never cursive, whereas minuscules are so commonly
connected by ligatures that the terms minuscule and cursive
are habitually used as synonymous, though they are not
properly so. In papyri the circumstances are quite different.
It is impossible to draw any distinction between uncial
and minuscule; and uncials, no less than minuscules, may

be written cursively. An uncial hand without ligatures is not necessarily earlier than one which has them; and, for non-literary purposes, hands of the most cursive character are found in the very earliest papyri yet discovered. The only classification which is of use in the study of papyri is that which has been stated above, into literary and non-literary hands; and it will be necessary, in the present treatment of the subject, to deal with them separately.

The material for the examination of these two branches is, however, of very different character. It is of the nature of a business document, such as a lease, a loan, or a receipt, that it should be accurately dated; it is only rarely that a literary work will have any precise indication of the same kind. The long series of dated documents, spoken of in the preceding pages, consists almost exclusively of non-literary hands. Consequently our knowledge of non-literary palaeography is far more exact than that which we have of literary papyri. A non-literary document must be written in a strange hand indeed if the doubts as to its period range over a hundred years, while those which are written in anything like the hand of a trained clerk can generally be placed approximately within the limits of a generation. But the dates assigned to literary MSS. have fluctuated over several centuries, and cannot even now be fixed with absolute precision. Yet it is just of these MSS. that it is most important to know the age. The precise date of a petition from the fellah Sarapion to the magistrate Hierax is a matter which concerns a few specialists alone; but the date of a MS. of Hyperides interests Greek scholars in general, and that of a copy of a Gospel, if one should be discovered, would be a matter of the gravest importance to theologians. The main object of the present essay is, consequently, to show how far science has progressed in this department; to examine the whole series of extant literary papyri; to show which can be dated accurately, and what are the

Non-literary palaeography less important but better known than literary.

probable dates of those which are still in doubt. Such an examination will have to deal with the subject practically *de novo*: and it is hoped that this may be done without the smallest appearance of disrespect for the eminent scholars and palaeographers who have previously assigned dates to the extant MSS. It is a case in which an access of fresh material justifies a new comer in revising the work of his betters; and the purpose of the following pages is rather to show where new facts have come to light than to add to the number of conjectural dicta.

Necessity of studying both.   In spite of the line which has been drawn between literary and non-literary papyri, it will not be possible to ignore the latter altogether in a study of the former. In the first place, the non-literary documents furnish us with certain broad criteria which are applicable to the literary documents. The main division, which will be made below, into Ptolemaic, Roman, and Byzantine periods, though more noticeable in the non-literary hands, is traceable also with the literary: and certain forms of letters are common to both. But in addition to this the evidence on which literary papyri are dated not unfrequently depends on our knowledge of non-literary palaeography. Titles, scholia, and corrections to a literary MS. are often written in non-literary hands, and so supply at least a *terminus ante quem* for the document in which they are found. In other cases, one side of the papyrus may contain literary writing, the other non-literary; a knowledge of the manner in which papyrus rolls were made and written determines which of these writings is the earlier, and a knowledge of non-literary palaeography gives us a date to guide us in estimating the age of the more valuable writing.

On all these grounds, then, it has seemed advisable not to limit this essay to the palaeography of literary papyri, but to include in it a sketch of non-literary writing as well. The latter will come first, because it is better known, and because the results of it are required for the examination of the less-known branch of the subject; but, both because it is better known and because it is less important, it may be treated more briefly. The more important literary papyri, on the other hand, will need individual examination; even small scraps of writing of this type have been noted as far as possible, since they will sometimes be found to throw valuable light upon the subject. But before proceeding to the palaeography, strictly so called, of papyri, it will be necessary to give some account of papyrus as a material for writing on,—the *Buchwesen* of the papyrus period, if a German word may be used where there is no exact equivalent in English. It will not be necessary to treat of this exhaustively, since it would be useless to reproduce what is already accessible in the recognized handbooks of the subject; but it will be useful to summarize what has been previously known, and to add the further information which is now available. Until recently, our knowledge of the subject has been almost entirely derived from the statements of ancient authors; but we are now able to test and interpret these by an examination of the extant papyrus MSS. which the sands of Egypt have restored to us.

CHAPTER II

PAPYRUS AS WRITING MATERIAL

Early use of papyrus in Egypt. THE use of papyrus in Egypt, the country of its production, goes back to an indefinite antiquity. The earliest extant specimen is a papyrus containing accounts of the reign of king Assa, whose date, according to a moderate estimate of Egyptian chronology, is about 3580–3536 B.C.[1]; while the earliest literary work which has come down to us (the Prisse Papyrus at Paris), although the copy which we have of it seems to have been written between 2700 and 2500 B.C., is stated to have been composed (and therefore originally written down) in the same reign. In technical execution the papyrus rolls of the Egyptian kingdom leave nothing to be desired, and there is no reason to suppose that the manner in which the material was prepared differed at all from that of Greek and Roman days. We have, however, to pass over a period of some three thousand years before reaching a date at which we can be sure that papyrus was in use among the Greeks. The statement of Herodotus[2] that the Ionians still applied the name of διφθέραι to books made of papyrus, because they had formerly used skins as writing material, shows at least that in his time the Greeks of Asia used papyrus, and that its use was not quite a new thing. It is safe, therefore, to assert that it was the writing material of the Greeks at the beginning of the fifth century; but how much further it can be

---

[1] Petrie, *History of Egypt*, i. 81.　　　　[2] Herod. v. 58.

carried back must remain doubtful. Formerly it was
held that the exclusiveness of the Egyptians in the matter
of foreign trade made it inadmissible to argue from the
practice of Egypt to that of Greece; but the proofs of
intercommunication have now increased so greatly that
this argument has lost force, and we must be prepared for
the possible appearance of evidence establishing the use
of papyrus by the Greeks at a much earlier period than
has hitherto been held probable.

With regard to the manner in which the writing material
was prepared from the plant, and the methods of writing
upon it, not much has to be added to what has long been
known on the subject and may be found in the recognized
authorities, notably Birt, Gardthausen, and Thompson.
The *locus classicus* on the preparation of the material is
a well-known passage in the elder Pliny (*N. H.* xiii.
11–13), the interpretation of which must be guided by the
conclusions derivable from the papyrus documents actually
in existence. The pith of the stem of the papyrus plant
was cut into thin strips, the width of which was of course
determined by the thickness of the stem, while their length
varied considerably, as will be shown below. These strips
(Lat. *philyrae*) were laid side by side to form a sheet.
Each sheet was composed of two layers, in the one of which
the strips ran horizontally, while in the other they were
perpendicular. The layers were attached to one another
by glue, moistened with water,—preferably, it would appear,
the turbid water of the Nile, which was supposed to add
strength to the glue[1]. The sheets thus made were pressed,
dried in the sun, and polished, so as to remove unevenness
in the surface; and they were then fit for use.

Preparation of the material.

---

[1] It is almost certain that this is the true sense of Pliny's 'turbidus
liquor vim glutinis praebet,' as held by Birt and Thompson. Apart from
the inherent improbability of the Nile water being able by itself to act as
glue, it may be added that traces of glue are visible in extant papyri.
The amount used was, no doubt, very small, so as to avoid affecting the
surface of the papyrus.

The size of the sheets (κολλήματα) varied according to the quality of the papyrus, only one rule being constant, that the height is greater than the breadth, when the sheet is held in the way in which it is meant to be used. Pliny (*l. c.*) gives the dimensions of the different qualities known in his time, and his figures were formerly supposed to apply to the height of the papyrus. This view is, however, invalidated by the testimony of the extant papyri, and Birt is no doubt right in referring Pliny's measurements to the width. The largest size named by Pliny (which was also the most valuable) is 13 *digiti*, or about 9½ inches; but papyri are extant which are as tall as 15½ inches, and from 10 to 12 is quite common in documents which make no pretence of special handsomeness or excellence. Further, we do not find that poor people necessarily use papyrus of very small height; while it is true that narrow sheets are often used for comparatively unimportant purposes. Thus, to quote cases in which only a single sheet is required for a document, the two most elaborately written petitions in the British Museum (Papp. CCCLIV and CLXXVII), which are addressed to the highest official in Egypt, measure respectively 8½ and 6¼ inches in width; while receipts, records of loans, and the like, are very commonly written on papyrus not more than three inches wide. An examination, moreover, of such long rolls as are at present known tends to confirm Birt's view that Pliny's dimensions refer to the width of the κολλήματα. The finest literary papyrus in existence, the British Museum *Odyssey* (Pap. CCLXXI) has κολλήματα of just over 9 inches in width; while in the Bacchylides papyrus, which is likewise a handsome roll, they vary between 8 and 9 inches. In the Herodas MS., which is small in height and unostentatiously written, they are only 6 inches in width. The papyrus of Hyperides *In Philippidem* and Demosthenes' Third Epistle, which is only 9¼ inches in height, has κολλήματα 7½ inches wide; while in a tax-register (Brit. Mus. Pap. CCLXVIII), which reaches the extraordinary height of 15½ inches, they are only 5 inches wide. The papyrus of the Ἀθηναίων Πολιτεία, which was originally intended merely for a farm-bailiff's accounts, has κολλήματα of 5 to 5½ inches in width ; and this is a very common size for non-literary documents.

For non-literary documents, such as letters, receipts, deeds of sale or lease, contracts, petitions, and the like, single sheets of papyrus could often be used : but for literary purposes a number of sheets were united to form a roll. According to the ordinary interpretation of Pliny's words, not more than twenty sheets went to a roll ; but this statement is not confirmed by the extant papyri. Twenty **Forma-** sheets of even the widest kind (9½ inches each, as stated in **tion of** the last paragraph) would only give a roll of 16 feet ; and **papyrus.** this length was certainly often exceeded. Egyptian papyri **rolls of** sometimes run to enormous lengths, in one case to as much as 144 feet ; but these need not be taken into account. They were for show rather than for use,—*éditions de luxe* which the owner proposed to take with him to the next world, where he might have strength to grapple with them, but which he certainly did not want to read on earth. But even Greek papyri, though they do not approach these dimensions, often exceed the limit which the text of Pliny appears to assign to them. From 20 to 30 feet may be **Their** taken as a full size, the higher limit being rarely, if ever, **length.** exceeded [1]. The largest papyrus of Hyperides (Brit. Mus. Papp. CVIII, CXV), containing the orations against Demosthenes and in defence of Lycophron and Euxenippus, must have measured, when complete, about 28 feet ; a MS. of the last two books of the *Iliad* (Brit. Mus. Pap. CXXVIII), about 25 feet ; and that of the *Mimes* of Herodas (Brit. Mus. Pap. CXXXV) perhaps about the same. On the other

---

[1] The papyrus containing the Revenue Laws of Ptolemy Philadelphus, obtained by Mr. Petrie and now in the Bodleian, measures 42 feet ; but this is composed of several distinct documents attached to one another, and, moreover, is not a literary work.

hand, a papyrus containing the *De Pace* of Isocrates (Brit. Mus. Pap. CXXXII) measures 14 feet; while that of Hyperides *In Athenogenem* cannot have measured, when complete, more than about 7 feet. The true interpretation of Pliny's statement, no doubt, is that in practice the sellers of papyrus kept and sold it in lengths (Pliny's *scapi*) consisting of twenty sheets. Egyptian rolls have been observed in which the number 20 is marked at the end of each twentieth κόλλημα, and this no doubt indicates the end of each length of papyrus as purchased by the author from the stationer [1]. But the author was no more limited by this fact than the modern writer is limited by the fact that he purchases his foolscap by the quire or the packet. He could join one length of papyrus on to another, and when he had finished his work he could cut off whatever papyrus was left blank. It is not in the least probable *a priori* that there was any hard and fast rule fixing the length of a papyrus book, nor do the facts disclosed by the extant MSS. authorize such a supposition.

Their height.

The height of a papyrus varies considerably, but the average may be taken to be from 9 to 11 inches. The tallest at present known is a tax-register in the British Museum (Pap. CCLXVIII), which measures 15¼ inches. Three census rolls (Papp. CCLVII–CCLIX) measure 13½ inches; and about 11 inches is quite a common height. Literary MSS. are generally rather smaller. Of those mentioned above, the principal Hyperides MS. measures 12 inches, the Isocrates 11 inches, the Homer 9¾ inches; while the Herodas (which must be regarded as a kind of pocket volume, such as volumes of poetry often are in modern times) is only 5 inches in height. The Louvre Hyperides measures 9 inches, the British Museum MS. of the same author's *Oration against Philippides* 9¼ inches; while the two most handsome literary papyri now extant, the British Museum *Odyssey* and Bacchylides,

---

[1] See Borchardt, *Zeitschrift für ägyptische Sprache* xxvii. 120, Wilcken, *Hermes*, xxviii. 167.

measure respectively 13 and 9¾ inches. To give the oldest extant examples, the Petrie *Phaedo* measures 8¼ inches in height, and the *Antiope* 8½ inches.

The writing was normally on that side of the papyrus *Recto and* on which the fibres lay horizontally (technically known *verso.* as the *recto*): and this is a rule of much importance; for when, as is frequently the case, a papyrus has been used on both sides, it is often only by this rule that it can be determined which writing is the earlier. It is therefore necessary to state the law somewhat precisely, following Wilcken, to whom the first formulation of it is due[1]. It is obvious on reflection that, by holding the papyrus differently, the fibres on *either* side can be made to lie horizontally, only in one case they are parallel to the height of the roll, and in the other to its length; but it does not follow that either side can be made the true *recto* at will. The true *recto* of the original sheets of papyrus out of which a roll is made is that side on which the shorter fibres are (probably because greater perfection and evenness could be secured with short fibres than with long ones); and the right way of holding such a sheet is to make these fibres lie horizontally (since thus the least obstacle is offered to the pen). Hence, when several such sheets are joined together, side by side, into a roll, all the fibres on the side which is primarily intended for writing will lie horizontally; and, conversely, the side of a roll on which the fibres run in the direction of the length of the roll is that which is primarily intended for writing. If writing is found running at right angles to the fibres, one of two things must have happened: either the scribe has written on the *verso* of the papyrus, or he is holding the *recto* in an unusual way. Examples of both are found. The *verso* is used when the *recto* already has writing upon it, or, *occasionally*, in the Ptolemaic period, without obvious cause;

[1] 'Recto oder Verso,' in *Hermes*, xxii. (1887).

C 2

never, so far as present experience goes, in Roman times, unless the *recto* has been previously used. Writing on the *recto*, but at right angles to the fibres, is found in a few Ptolemaic documents and in many of the Byzantine period. In the former cases the roll has been made up in an unusual manner, the sheets being joined together top and bottom, instead of side by side; so that the writing, though it is across the fibres, still runs parallel to the length of the roll. In the Byzantine documents, on the other hand, the roll is made in the ordinary way, and the writing runs parallel to the height of the roll, so that in reading the roll has to be unfolded from the top downwards, instead of sideways. This method is not applied, however, to works of literature.

The general rule,—invariable in the case of literary works, apparently invariable in non-literary works of the Roman period, and largely predominating in non-literary works of the Ptolemaic period,—is that the first writing on the papyrus is parallel to the fibres. The exceptions can easily be recognized after a little experience,—in the case of large papyri, by looking for the traces of the original sheets, and in the case of small scraps by noticing the greater smoothness of the true *recto*. It is rare to find a work continued from the *recto* to the *verso*. A long book of magical formulae (Brit. Mus. Pap. CXXI) and an account book (Pap. CCLXVI) are examples of this practice; but ordinarily if there is writing on the *verso* it is quite independent of that on the *recto*. It may safely be assumed that no MS. of a literary work intended for sale was ever written on the *verso*; though copies for private use might be, as in the case of the Ἀθηναίων Πολιτεία of Aristotle. In a long medical papyrus in the British Museum (Pap. CXXXVII), and in the Revenue Papyrus of Philadelphus, notes or additions to the main work are written on the *verso*; but neither of these is a literary work intended for sale.

Width of columns.   The normal method of arranging the writing on a papyrus roll was in columns, the lines of which run parallel to the length of the roll, as above described. For literary MSS. this method is invariable; the Byzantine documents mentioned above, in which the writing ran in one large column down the whole papyrus, in lines at right angles to its length, are of a non-literary character, generally wills or leases. The width of these columns (σελίδες) varied, but for literary MSS. intended for sale the length of a hexameter line may be taken as determining the extreme width. This, in a hand of good size, implies a width of about 5 inches, besides the margins, which might be as much as $1\frac{1}{2}$ inches between the columns and 2 or 3 inches at the top and bottom [1]. In prose works, so far as our present knowledge goes, the width of the columns is generally much less. The widest are found in the Louvre papyrus of Hyperides, which measure about $3\frac{1}{2}$ inches; the narrowest in one of the British Museum papyri of the same author (the *In Philippidem*), which are barely half that width ($1\frac{3}{4}$ inches). The large Hyperides papyrus has columns 2 inches wide, while those of the British Museum Isocrates measure a little less than 3 inches. The only literary papyrus in which these dimensions are exceeded is that of Aristotle's Ἀθηναίων Πολιτεία, which has one column measuring as much as 11 inches wide, while several others are 5 or 6 inches; but this does not constitute a real exception, since the MS. is not written in a literary hand, nor intended for publication [2]. In non-literary papyri much

---

[1] These are the measurements in the British Museum papyrus fragment of the *Odyssey* (Pap. cclxxi, which may be regarded as the handsomest literary papyrus at present extant. In another well-written Homer MS. (Brit. Mus. Pap. cxxviii the figures are slightly less.

[2] It has sometimes been supposed that the σελίδες correspond with the κολλήματα, that is, that the writing was not allowed to cross the junctions between the sheets of which the papyrus roll was composed; but this is not borne out by the facts. In the best-written MSS. (such as the British Museum *Odyssey* and Bacchylides), no less than in the worst, the writing frequently crosses the junctions.

greater widths are sometimes found, their dimensions being in fact determined by their contents. In census-lists and some kinds of accounts the scribe preferred to get each entry into a single line; hence in a document of the former class (Brit. Mus. Pap. CCLX) some of the columns are 10 inches in width, while in a tax-register (Pap. CXIX) they are as much as 12 and 13 inches. Ptolemaic scribes had a fondness for writing such things as wills, leases, loans, and the like, in one or two very broad columns. To give two instances only, Brit. Mus. Pap. DCLXXV is written in a single column measuring 15 inches in width and only 6 inches in height; while Pap. DCXXIII is in two columns of 13 inches, besides a short abstract of the contents in a separate column. These, however, are only examples of a temporary fashion, which must have been found inconvenient in practical use, and could never have been adopted for any composition which was likely to be often read.

Titles. The mutilation of nearly every literary papyrus which has come down to us renders it difficult to lay down any very certain rule as to the methods commonly employed by the ancients to indicate the contents of a roll. It was certainly not unusual to inscribe the title of a work at the end of it, as is the case with the largest Hyperides MS., the British Museum Isocrates, and several of the papyri of Homer; but it was not invariably done, for it is not found in the MS. of the *In Philippidem*, the end of which is perfect, nor yet in the Aristotle papyrus. It is also obvious that it would have been extremely incon-venient to have to unroll the whole of a volume in order to see what its contents were. It was usual to leave a blank column at the beginning of a roll, as is found in the papyri of Aristotle and Herodas and in the Harris Homer; but in none of these cases is the title written on it; and in the great Hyperides MS., where the title was so written, it is by a different and apparently later hand. It seems

certain, therefore, that the ordinary way of indicating the
title of a work was by the σίλλυβος, or little strip of papyrus
or vellum attached to and projecting from the roll; and
these, though known from the references in ancient
authors to have existed [1], have in no case come down to us.

In all palaeographical works it is stated that the roll, Use of
when completed, was rolled on a stick (ὀμφαλός), ornamented rollers.
at the ends with projecting knobs or tips (κέρατα); and
the statements of Latin writers [2] leave no doubt that this
was the habitual practice in the case of their works.
The actual papyri which have come to light of recent
years make it necessary to modify this proposition. In
no case (except in dummy rolls manufactured for sale to
tourists) has a wooden roller been found; many of the
Herculaneum papyri had a central core of papyrus: some
burnt rolls brought from Egypt a few years ago had, in
some instances, a reed or quill in the middle; but as a rule
there is no trace of any roller at all. This fact is perfectly
intelligible in itself, and may quite well be reconciled with
the statements of the Latin poets. Papyrus was not
originally the brittle material which, from its appearance
after the lapse of a score or so of centuries, one is apt to
imagine, and could quite easily be rolled upon itself; and
for this purpose, as well as to resist tearing, the ends of
the roll are often stiffened by an extra thickness of papyrus,
about an inch in breadth [3]. In the case of common copies
this was, no doubt, the regular practice; while the
handsomer books would be provided with wooden rollers
and all the other appurtenances of style and luxury. The
distinction would be very much the same as that between
cloth-bound and paper-covered copies at the present day.
The Latin poets are speaking of the dainty copies of their
works which were to be seen in the bookshops and salons

[1] Thompson, *Greek and Latin Palaeography*, p. 57.   [2] *Ib.* p. 56.
[3] A good instance may be found in the Harris Homer (Brit. Mus.
Pap. cvii .

of the capital ; while the papyri which have come down
to us are generally from the houses and tombs of obscure
provincials in Upper and Middle Egypt.

On the further details of book production in the papyrus
period there is nothing new to be recorded. The allusions
in the Latin poets provide us with all we know as to
the φαινόλης, or wrapper, with which the roll might be
covered, the cedar-oil by which it might be protected
against insects, the chest (resembling a bucket, to judge
from extant representations) in which it was kept. These
details have already been gathered together in the recognized
handbooks of palaeography, and it seems useless to repeat
them here. The discoveries of actual papyrus rolls have
added nothing to our knowledge on such points.

Papyrus
codices.

The description of a papyrus book, which has been
given in the preceding pages, applies to nearly the whole
of the papyrus period, as at present known. But towards
the end of the period the codex, or modern book form,
is found coming into existence side by side with the
traditional roll form. The origin of the codex is, no
doubt, to be found in the sets of wax tablets which were
in use for note-books at least as far back as the first
century B.C., and probably much earlier. These tablets,
consisting of wax laid upon wood and surrounded by
raised wooden rims, were bound together by strings or
leather bands passing through holes bored in the rim
of one of the longer sides of each tablet, so as to form
something in the shape of a modern book. It was not
until several centuries later, however, that this shape
was adopted for literary compositions. As Sir E. Thompson
has pointed out [1], the cause of the final victory of this
form was that it was possible to include so much more
matter in a codex than in a roll ; and the requirements
of the churchmen and the lawyers agreed in giving it

[1] *Greek and Latin Palaeography*, p. 61.

the preference. A single Gospel was as much as a papyrus roll could contain, while a vellum codex could include the entire Bible; and the great legal collections of the reign of Justinian would have needed a whole library of rolls. For a short time the experiment was tried of making papyrus codices, the papyrus being cut up into leaves and fastened together by strings or leather bands. It does not appear, however, that the experiment gave satisfaction, for very few instances of such papyrus books have come down to us. A MS. of the second, third, and fourth books of the *Iliad*, formerly assigned to the fifth century, but, for reasons which will be given subsequently, more probably belonging to the third (Brit. Mus. Pap. CXXVI); a page of Menander at Geneva, probably of somewhat later date; the Berlin fragments of the 'Αθηναίων Πολιτεία, probably of the fourth century; a copy of the prophet Zechariah, of the sixth or seventh century, exhibited at the London Oriental Congress in 1892, and last seen in a dealer's shop in Vienna; some leaves of a Psalter of the same date, in the British Museum (Pap. XXXVI); small portions of two Hesiod MSS. at Paris and Vienna, of the fourth or fifth century; a magical papyrus in the British Museum (Pap. XLVI), and another in the Bibliothèque Nationale (MS. Suppl. grec. 574), both of the fourth century: these are the principal examples of Greek papyrus codices at present known to exist. There are Coptic volumes of this kind of much greater size, resembling the large vellum quartos and folios, but there is no sign that these were ever adopted for Greek literature. The rise of the codex was accompanied by the rise of vellum, and the papyrus period, so far as Greek literary works are concerned, was then coming to an end.

Before beginning the history of Greek writing on papyrus, and of the evolution of the written characters individually, it will be as well to dispose of some subsidiary

Accents, breath- ings, and stops.

matters, such as punctuation, accentuation, and the use of breathings. In all these respects papyrus MSS. are in a very elementary stage. None has a full equipment of stops, accents, and breathings; many have none at all. Further, the use of them does not follow regular laws of development; or rather, the materials now available do not allow us to ascertain any law. The probability is that, the higher the quality of a MS., the fuller is its equipment in these subsidiary guides to intelligence. Non-literary MSS. very rarely have any of them. Accents are not found in them at all; breathings in extreme rarity; and only a few have some show of division of sentences. In the lower classes of literary MSS. an occasional accent is found, probably when there was some likelihood of a mistake as to the meaning of a word; breathings are equally rare; and only important breaks in the sense are indicated by punctuation-marks or blank spaces. Of the highest class of MSS., those which were intended for sale or for preservation in large libraries, there are very few extant specimens: but there is some indication that accents, breathings, and punctuation-marks were more freely used in them, though without any approach to the completeness of later usage.

Separation of words. The most elementary form of assistance to the reader consists of the separation of words from one another. Where this exists, it is not usually very difficult to determine for oneself the pauses in the sense. But, perhaps because it is so elementary, it is the last form of assistance to be given in Greek MSS. There is sometimes an approximation to it in non-literary papyri, where, the text being written cursively, the writer not unnaturally lifts his pen oftener at the end of a word than elsewhere; but this is so irregular and incomplete as to furnish very little help. In literary papyri the separation of words is almost wholly wanting; perhaps the only example of it is in a short grammatical treatise, bearing the name of Tryphon, written not earlier than the fourth century on some blank pages in a MS. of Homer in the British Museum (Pap. cxxvi). In other MSS. the nearest approach to such a practice is the use of a dot, above the line, to indicate the true word-division in cases where the reader might easily make a mistake at first sight. Thus Brit. Mus. Pap. cxxvi gives, in Homer, *Il.* iii. 379, ἄψ · ἐπόρουσε [1]: and a similar system is found in the Marseilles papyrus of Isocrates. It is however rare, the dot, if used at all, being generally required to separate sentences rather than words. A comma below the line is also found occasionally for this purpose (*e. g.* Bacch. xvii. 102).

Punctuation in the ordinary sense of the word, or the indication of a break in the sense, is more common, though still only sporadic. The earliest system would seem to be that of leaving a slight space in the text, and placing a short horizontal stroke (παράγραφος, more rarely παραγραφή) below the beginning of the line in which the break occurs [2]. This use of the παράγραφος is mentioned by Aristotle [3], and is found in some of the earliest extant papyri. Thus in the fragment of the *Antiope* among the Petrie papyri (third century B. C.) it is used to indicate the end of each actor's speech ; and similarly, along with the double dots mentioned below, in the Petrie *Phaedo* fragment. It is found also in the Louvre Hyperides (second century B. C.), the British Museum MSS. of the same author (first century B. C. and first century after Christ), the Herodas (first or second century), and in several other MSS. In the Bacchylides MS. (first century B. C.) it marks the end of each strophe, antistrophe, and epode. Mistakes are sometimes made by transcribers in placing the παράγραφος, but its proper place is *below* the line in which the pause occurs. It marks the end, not the beginning, of a sentence.

Punctuation.

[1] *Classical Texts from Papyri in the British Museum*, p. 82.

[2] Spaces in the text, without *paragraphi*, are found in some literary papyri (*e. g.* the Herodas MS. sometimes), and not unfrequently in non-literary papyri, especially those of a legal nature. [3] *Rhet.* 3. 8. 1409 a. 20.

The use of dots for the same purpose is equally old. In the Petrie *Phaedo*, which is of the same age as the *Antiope*, and in the Vienna papyrus containing the 'Curse of Artemisia,' which may be older, a double dot resembling a colon is used to separate sentences. This occurs also in the *Erotic Fragment* found and published by Mr. Grenfell (now Pap. DCV in the British Museum), which is of the second century B.C.; but it is not common. The single dot, generally placed well above the line, is common. It is used freely, and almost regularly, in the Bacchylides papyrus, occasionally in the British Museum *Odyssey* MS. (Pap. CXXXI); and it has been added by later hands to the three earliest papyri of the *Iliad* in the British Museum (Papp. CVII, CXIV, CXXVIII). The use of the dot is known to have been systematized by the Alexandrian critics (traditionally by Aristophanes of Byzantium), and different values were assigned to it according as it stood above the line (a full stop), in the middle of the line (a comma), or on the line (a semicolon): but this system cannot be traced in the extant papyri, where the dot is generally above the line (practically never on it), and is used to indicate minor pauses, such as a semicolon or even a comma, quite as much as for a full stop.

Accentuation. Accentuation is rarer than punctuation in Greek papyri, and quite as fluctuating in its appearance. It is not found at all in non-literary documents, and in literary MSS. its use is sporadic. It does not appear in the Petrie papyri of the third century B.C., nor in the Louvre Hyperides of the following century. The earliest example of the use of accents is in the Bacchylides MS., where they are also more plentiful than in any other papyrus; and it is worth noticing that this is likewise one of the most carefully written papyri in existence, and is therefore probably something more than a copy for private use. Accents are also somewhat freely used in the Alcman fragment in the Louvre, which is probably of the latter part of the first century B.C.: and more scantily in the British Museum *Odyssey* papyrus (early first century) and two fragments of the *Iliad* in the Louvre (first and second centuries). The Harris and Bankes papyri of the *Iliad*, which will be found placed below in the first and second centuries, though the former has hitherto been held to be older, have many accents, but not by the first hand, so that their date is uncertain; and the same is true of the British Museum papyrus of the last two books of the *Iliad* (Pap. CXXVIII). In the much later MS. of *Iliad* ii-iv (Pap. CXXVI), which is probably of the third century, the accents are by the first hand. None of the Hyperides MSS. in the British Museum has accents, and the Herodas only a few isolated examples.

It is thus clear that accents are not to be looked for in papyri with any confidence, and are never used to the full extent that has since become customary. Even fairly well written MSS., such as the Petrie *Phaedo* and *Antiope*, the Hyperides papyri, and the British Museum *Iliads* (with the exception of the latest), have none by the first hand. It is perhaps significant that the only two texts (earlier than the third century) in which they are at all largely used are both of them lyric poets. It may well be that in Bacchylides and Alcman the scribes felt that the reader required more assistance than in Homer or Hyperides. In these MSS. accents are especially applied to the longer words, and particularly to compounds, which are somewhat misleading to the eye. Prepositions, articles, pronouns, and adverbs very rarely have them, unless there is something abnormal about them, as when a preposition follows its case (*e. g.* κράτος ὕπερ in Bacch. xviii. 51). In the case of diphthongs, the acute accent is generally on the first letter, and the circumflex over both, contrary to modern usage. In the Bacchylides the accent is never placed on the final syllable in oxytone words, but the preceding syllables have the grave accent; *e. g.* πὰγκρὰτης, κρὰτος.

The explanation of this is to be found in the original theory of accentuation, according to which every syllable has an accent, but only one in each word is acute, the rest being grave; e.g. ἄνθρωπος, καρδία, should be written ἄνθρὼπὸς, κὰρδίὰ. In practice the grave accents were omitted; but why they were revived, to the exclusion of the acute, in oxytone words, is not clear. Traces of the same practice are found in the Harris and Bankes Homers: e.g. ἐλων, φρὲσιν, and (in a proparoxytone word) ἐπὲσσεύοντο [1].

The general principle governing the use of accentuation in the papyrus period would seem to be that accents were only inserted if the scribe felt they were wanted as an aid to reading, and (so far as yet appears) solely in texts of the poets. Just at the end of the period they came into more general use, and were sometimes supplied to previously existing MSS.; but at this stage the supersession of papyrus by vellum came about, and accents once more disappeared from Greek texts for some hundreds of years.

Breath-ings. Much the same may be said of breathings, which are usually found in the same MSS. as accents. No papyrus is early enough to show the letter Η in its original use as an aspirate; but the two halves of this letter, ⊢ and ⊣, indicating the rough and smooth breathings respectively, are found in the Bacchylides MS. (though not uniformly), and in a few instances in the British Museum *Odyssey* papyrus. The more usual forms, however, both in these MSS. and elsewhere, are ⌐ and ⌐, or ⌐ and ⌐. The rounded breathing is not found in papyri, though the inverted comma (') is used as a mark of elision in the Bacchylides and other MSS. As with the accents, breathings are only used intermittently, when the scribe thought them necessary in order to avoid confusion or mistake.

Other marks. Marks of diaeresis (··) are often used over ι and υ, especially at the beginnings of words. They are found in non-literary as well as literary papyri.

---

[1] *Catalogue of Ancient MSS. (Greek) in the British Museum*, pp. 1, 6.

Dots are sometimes placed over letters, to indicate that they are cancelled. This is especially found in the large Hyperides MS. and the Herodas; elsewhere it is more usual to draw the pen through the cancelled letters. Corrections are normally written between the lines, above the words for which they are to be substituted; occasionally (*e. g.* in the Aristotle papyrus) they are inclosed between two dots, but this is unessential. Omitted lines are supplied in the upper and lower margins, with a mark at the place where they are to be inserted (*e. g.* Bacchylides and Herodas MSS.). The margins (lateral as well as upper and lower) are also the place for scholia (*e. g. Odyssey* papyrus, Brit. Mus. Pap. cxxviii, etc.).

Other marks, such as 7 or = to fill up blank spaces at the end of a line (Brit. Mus. Hyperides MSS.), a hyphen (‿) under compounds to show that they are single words (Bacchylides MS.), and the like, are apparently due to the fancies of the individual scribes, and explain themselves. Two of the critical marks used by Aristarchus to indicate spurious or repeated lines in Homer (the διπλῆ, ≻, and asterisk) are found in Brit. Mus. Pap. cxxviii (*Il.* xxiii, xxiv), the Oxford papyrus of *Il.* ii, and perhaps in Brit. Mus. Pap. cclxxi (*Od.* iii); but they are not fully inserted in any of these MSS.

A brief mention should be made of the rules for the division of words in Greek papyri, because the point is often of importance in the restoration of mutilated texts. If a word has to be divided at the end of a line, the rule is that the division should be made after a vowel, except in the case of doubled consonants, where it is made after the first consonant, or where the first of two or more consonants is a liquid or nasal, when it is divided from the others. Thus, in the course of a few lines of Hyperides *Pro Euxenippo* the following examples occur: ἔλε|γον, δή|μῳ, δι|καστήριον, δικα|σταί, ἄλ|λοι, ἔχον|τες. In the case

Division of words at end of line.

of words compounded with a preposition, the division
is most commonly made after the preposition; but not
unfrequently the normal tendency to make the break at
a vowel prevails. Thus we have εἰσ|αγγελλομένων, προσ|-
ῆκεν, but also εἰ|σαγγελίας, ὑ|πελάμβανες. The same tendency
is seen at its strongest in such divisions as ταῦ|τ', οὐ|κ,
κα'θ', which are far from uncommon. Breaches of these
rules are practically unknown in literary papyri [1]. The
only point in which variation is admitted is in combina-
tions of σ with another consonant, some scribes making the
division before the σ, and some after it. Thus in some
MSS. we find δικασ|ταί, and in others δικα|σταί. Occasionally
the same scribe will fluctuate in his practice in this respect.
With regard to non-literary documents it is dangerous to
assert an universal negative, on account of their great
number: but a special search through a considerable body
of them has failed to find a single example of a division
contrary to the principles above stated, and it is at least
clear that the rule is so generally observed that any breach
of it must be regarded as quite exceptional.

Abbrevia-
tions. Abbreviations are not found in well-written literary
papyri, with the exception of two Psalters (late third
and seventh centuries), in which the common compendia
for κύριος, θεός, κ.τ.λ., occur. In literary texts written in
cursive hands, however, they are found somewhat largely
in use. The chief example is the papyrus of the Ἀθηναίων
Πολιτεία, where two of the writers whose hands occur in
the MS. use twenty-five abbreviations for common words
such as articles, prepositions, καί, οὖν, εἶναι, κ.τ.λ. The large
medical papyrus in the British Museum (Pap. CXXXVII)
uses nearly as many, some, but not all, being identical.
One of the Herculaneum papyri of Philodemus (Pap. 157–

---

[1] The contrary is stated by Wattenbach (*Griech. Paläogr.* 3rd ed., pp. 15,
118', who characterizes the division of words in Egyptian papyri as 'ganz
regellos'; but this is quite a misconception.

152) has a few abbreviations [1], and there are also a few
in the scholia to the Alcman fragment, and in a collection
of rhetorical exercises in the British Museum (Pap. CCLVI
verso); but these exhaust the list of literary texts containing such symbols. In non-literary papyri, as might be
expected, they are more freely used, especially in accounts
and receipts. Still more common, however, than the use
of symbols to denote the terminations of words (κ', τ', and
the like) is the practice of abbreviating words by the
simple omission of terminations. This is found in the
Aristotle and medical MSS. above-mentioned, but is especially frequent in non-literary documents. The regular
system of such abbreviations is to omit the latter part
of the word, and to elevate the last letter remaining above
the line, or else to draw a line over it as a mark of
abbreviation; thus either $\pi\rho o^\kappa$ or $\pi\rho o\bar{\kappa}$ may stand for
$\pi\rho \acute{o}\kappa\epsilon\iota\tau\alpha\iota$. Abbreviations such as these explain themselves,
and do not admit of tabulation; but a list of symbols used
in abbreviations is given in an appendix. Contraction,
in the sense of the omission of the middle portion of
words, such as occurs in mediaeval Latin MSS and in
modern letters, is not found in Greek papyri [2].

The study of tachygraphy is too special a subject to
be dealt with here, and the explanation of the few extant
specimens of it on papyrus is still obscure [3]. No long
document in this style of writing has yet been discovered;
and though several small examples are said to be in the
Rainer collection at Vienna, they have hitherto been only
imperfectly published, and the explanation of their systems
has, in most cases, still to be given.

Tachy-graphy.

[1] Scott, *Fragmenta Herculanensia*, p. 98.

[2] A supposed instance to the contrary has been pointed out in Grenfell's
Greek Papyri, i. 24 (now Brit. Mus. Pap. DCXX), l. 6, where $\beta\alpha\sigma\sigma\eta s$ is given
as the reading of the MS. for $\beta\alpha\sigma\iota\lambda\acute{\iota}\sigma\sigma\eta s$. In reality the word is not con-
tracted at all, but simply written in a very cursive fashion.

[3] See (in addition to the older literature) Gitlbauer, *Die Drei Systeme der
griechischen Tachygraphie* (1894), and Wessely, *Ein System altgriechischer Tachygraphie* (1895), both in the *Denkschriften* of the Vienna Academy.

CHAPTER III.

The three periods of papyrus-palaeo-graphy.

THE history of Greek writing upon papyrus has three well-marked periods, the distinction between which is the foundation of all palaeographical knowledge of the subject. These periods correspond to the three political administrations by which the country was successively governed after the extinction of the native Egyptian empire. From 323 to 30 B.C. it was under the sway of the Ptolemies; from the conquest by Augustus to the re-organization of the empire by Diocletian, it was administered from Rome; and from the time of Diocletian to that of the Arab conquest in A.D. 640, it was a part of the eastern or Byzantine division of the Roman world. The changes in the prevalent type of handwriting curiously reflect the changes in the administration; and the classification of hands as Ptolemaic, Roman, or Byzantine is not merely conventional, but corresponds to real differences of character which can be made obvious to the most untrained eye.

It is, however, only of the non-literary hands that this proposition is true in so extreme a form; and the reason for this does not seem difficult to imagine. The pattern

---

[1] In parts of this chapter I have made use of my own treatment of the same subject in the Introduction to the first volume of the *Catalogue of Greek Papyri in the British Museum* (1893), but with amplifications and modifications according to the new material which has come to light since that was written. The book is, of course, one which will have come into the hands of few but specialists.

for the non-literary hands—not merely for those of professional clerks, but, through them, also for those of private persons—was set by the government officials, and varied according as the higher members of that class came from Alexandria, Rome, or Constantinople. The literary class, on the other hand, had no direct dependence on the political capital. If they were dependent on any outside influence, it was that of Greece, the fountain from which their higher inspirations were drawn, and with which they were connected by a strong and continuous tradition. Only gradually, and at some considerable distance, were the fashions of literary manuscripts affected by the contemporary varieties in every-day writing ; and it requires a little practice to see where the characteristics of non-literary papyri manifest themselves in the literary hands of the same period. That they do manifest themselves, however, will, it is hoped, be shown in the following chapters ; and the full and certain knowledge which we now have of non-literary palaeography goes far to lay a firm foundation for the more interesting and important branch of the subject which deals with the literary manuscripts.

The beginning of the history of papyrus-palaeography is fixed, for the present, by the discoveries of Mr. Flinders Petrie in 1889, when he extricated a mass of documents of the third century B. C. from a number of mummy-cases found at Gurob. The mummy-cases, instead of being of wood, were made of a kind of papier-mâché, the material being papyrus, coated over with plaster. The papyri were evidently nothing but the produce of the waste-paper baskets of the period, torn, cut, pasted together just as they came, defaced by the plaster and mutilated by rough handling; yet these same rubbish-heaps, patiently sorted and set in order by Prof. Mahaffy, are now the foundation of our knowledge of Greek palaeography. To these must be added the great Revenue Papyrus of Ptolemy Philadelphus, the greater

I. Ptolemaic period : the Petrie papyri.

D 2

part of which was acquired by Mr. Petrie in 1893, and the rest by Mr. Grenfell in the following year; but this only confirmed by additional examples the knowledge which the Petrie papyri had already established. Before Mr. Petrie's discovery there were, it is true, a few documents of the third century already extant in some of the great European libraries; but they were undated, and there was nothing to show their real age, which, in consequence, was generally underrated. About the age of the Petrie papyri there could be no doubt. Many of them were wills, petitions, and similar documents, bearing precise and indisputable dates in the reigns of the earliest Ptolemies (with the exception of Ptolemy Soter, the founder of the dynasty). The earliest definite date was in the year 270 B.C., the latest was in 186 B.C.; and there was no reasonable doubt that the great mass of undated documents lay between these limits, and that nearly all of them belonged to the third century B.C. Subsequent examination of the previously undeciphered fragments of the collection has, indeed, revealed one or two later dates, but has done nothing whatever to shake the general conclusion. From some seventy definitely dated documents in a great variety of hands, and from many scores of undated documents of the same period, we have ample means for estimating the character of Greek writing in Egypt—and specifically in the Fayyum—in the third century B.C.[1]

The first characteristic which strikes the eye in the writing of this period as a whole is its freedom and breadth. The style is light and flowing, strokes are free and curved, without being necessarily careless and ill-

[1] The descriptions which follow will be more intelligible if read with the series of facsimiles published by the Palaeographical Society, or the atlases accompanying the *Catalogue of Greek Papyri in the British Museum* (vols. i. and ii.. A very useful table of alphabets is given in Sir E. M. Thompson's *Handbook of Greek and Latin Palaeography*. The facsimiles given in the present volume, though fairly characteristic of their respective periods, are too few to exhaust all their varieties.

OFFICIAL LETTER—B.C. 242.

formed. Such hands, indeed, there are among them, in which cursiveness borders closely on illegibility; but the characteristic hand of the period is graceful and easy, showing at once a full command of the pen and a plentiful supply of papyrus. Letters such as M, Π, T have an almost excessive breadth in their horizontal strokes, and it needs a good-sized piece of papyrus to contain any appreciable quantity of writing. The Roman style is altogether smaller and more compact; that of the Byzantine period, though its letters are sometimes quite as large, is squarer, generally heavier, and shows more signs of deliberate care and conscious style. The Ptolemaic scribe wrote freely and often well, but without self-consciousness.

Another characteristic of Ptolemaic writing is the appearance, which it generally presents, of a horizontal line along the *top* of the letters. Most of the letters are shallow, and the horizontal strokes in them are formed near the top, while the perpendicular strokes project very little above the line and are carried far down below it. Thus M (a very characteristic letter in Ptolemaic hands) is generally formed of two perpendicular strokes, stretching well below the line, and united at their tops by a horizontal stroke with a very shallow curve. This feature is not found in either of the later periods, and indeed is more characteristic of the third century B. C. than of the later Ptolemaic age. Good writing of the Roman period has few prominently projecting letters at all; while the Byzantine hand, though marked by very long upright strokes, presents the appearance of an even line along the *bottom* of the writing, with projections bristling along the top.

The specimen shown in the accompanying facsimile (Plate I) is part of a letter of the year 242 B. C., and is a fairly characteristic example of the hand of this period. The completely cursive character of the writing is obvious at first sight, as well as the line along the tops of the letters described above. There are no signs of difficulty

or of unfamiliarity with the use of the pen. It is clear
that writing of such freedom and ease has a long history
behind it.

The third
century
B. C.

With regard to the forms of individual letters in documents of the third century, the most characteristic are ⲇ,
M, N, Π, T, Y, and ⲱ. ⲇ is often found, especially in the
less carefully written papyri, in the shape of a simple wedge
(∠, ∠); M, as described above, has a very shallow curve
between two nearly perpendicular strokes (Π); N frequently
has its final upright stroke carried up far above the line
(ᴊ, ᴕ); Π is either broad and low or rounded into an almost
semi-circular curve; T is almost always without the right-
hand portion of its cross-bar, being written with a single
stroke of the pen; Y generally has a large loop, carried
further to the left than the right (ⲍ); while ⲱ habitually
has its second loop represented by an almost or quite
straight line. These characteristics are, no doubt, most
apparent when the writing is least careful and formal; but
few documents of the period are without some of them.
There is very little modulation of fine and thick strokes.
As a rule the lines, though not heavy, are thick and black;
and there is much unevenness in the size of the letters.
Hence the writing of this century, free and flowing though
it often is, is rarely very ornamental. The enthusiasm
of the first discoverers led them at times to speak too
highly in its praise; but an unprejudiced comparison of it
with the hands of the succeeding centuries will show that,
at least in the specimens hitherto known, it is inferior
in regularity and handsomeness.

Early
second
century
B. C.

For the first half of the second century B. C. our material
is mainly drawn from the papyri found in the Serapeum
at Memphis, many of which were written by a single scribe,
a Greek recluse in the Serapeum named Ptolemy. There
are, however, several examples of other hands, which justify
us in using this group of documents as fair evidence for
their period. As a whole, these hands are larger, more

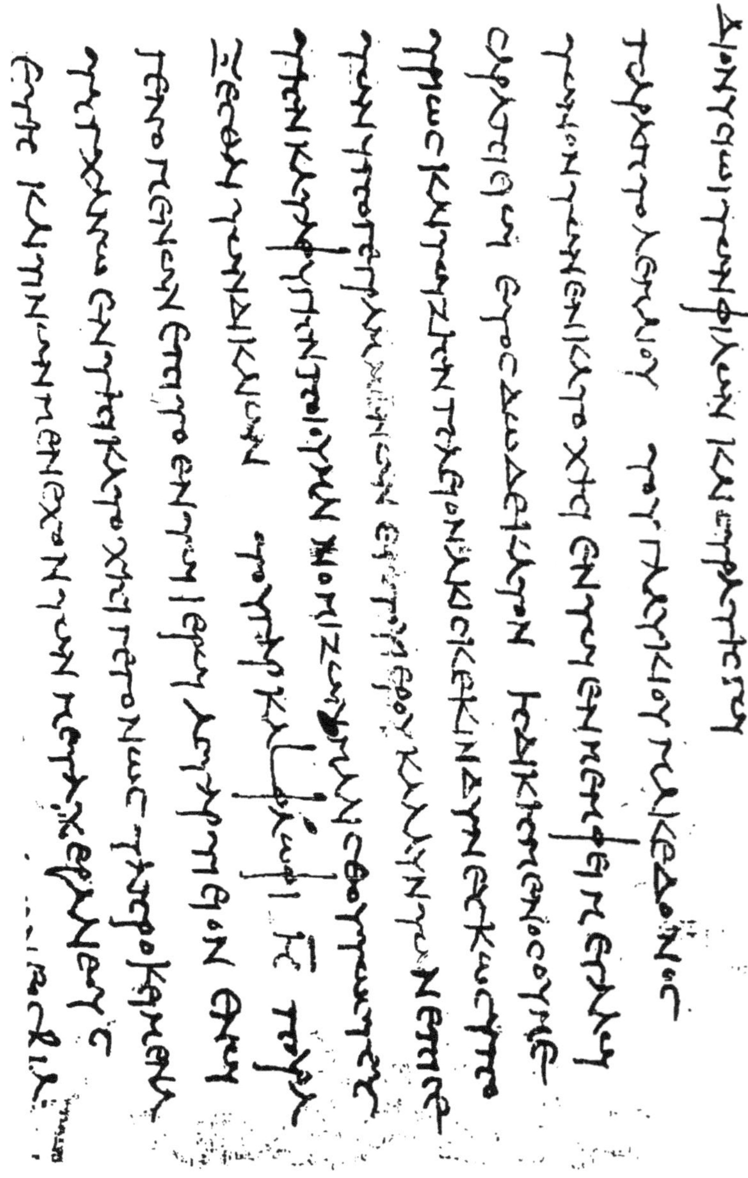

regular, and less cursive than those of the preceding century. The letters have more of an uncial form, and are only linked together in groups of two and three. The better-written specimens are clear and regular, and sometimes even handsome. The appearance of a line along the top of the writing ceases almost entirely. Of the individual letters, the wedge-shaped A is rarely found after the third century; it is of the minuscule type, but the loop is sometimes represented by a mere straight line ($\frown$); M is still sometimes marked by its shallow depression, but oftener the middle loop is carried lower down and bent to an angle in the centre, so as to resemble a rough uncial M; B is very large, often extending both above and below the line; the first stroke of H is higher than the second, which rarely rises above the cross-bar, and is linked to the following letter by a horizontal stroke from the top; K is generally large, especially the upright stroke; N is sometimes of the third century pattern, with the last stroke rising above the line, but is oftener of the normal uncial shape; T has acquired the right-hand portion of its crossbar, though it is still often written without lifting the pen, by making the cross-bar first and then drawing the pen backwards and downwards; at other times the first half of the cross-bar and the down-stroke are formed together, as in the third century, and the second half of the cross-bar is added separately, being often attached to the succeeding letter; Y is not unlike T, being made by forming a wide, shallow curve, and then drawing the pen backwards and downwards; ω is generally of the ordinary minuscule type, though the second loop is still sometimes clipped.

The facsimile shown in Plate II (from Brit. Mus. Pap. XLIV) is, perhaps, the best specimen of calligraphy among the Serapeum papyri, though another (Pap. XXIV) is in a larger and bolder hand. It is a petition from the above-mentioned Ptolemy, son of Glaucias, for protection and redress on account of an assault made upon him by some of the

Egyptian attendants in the temple. It is dated in the year 161 B. C., and is written in a clear, regular, and notably well-formed hand—certainly not Ptolemy's own, but probably that of a professional scribe.

The second half of this century was almost a blank, so far as palaeography was concerned, until the purchase by Messrs. Grenfell and Hogarth, in 1894–5 and 1895–6, of a large number of dated documents in very good condition, which cover this period very fully and extend into the following century. In these papyri, which come from the neighbourhood of Thebes, we see the revival of a fully cursive style of writing; or perhaps it would be more true to say that they justify the belief that a more cursive style than that of the Serapeum documents was in existence during the earlier half of the century, as it had been in the third century. The cursive hand of 150–100 B. C. (see Plate III) is, however, quite distinct from that of 270–200 B. C. It is smaller, better formed, and more ornamental; with quite as much liberty as its predecessor, but less licence. The scribes have learnt to be regular without being stiff, and the sizes of the letters are better proportioned to one another. It is the best period of the Ptolemaic cursive, and comes just before its decline and disappearance. At the same time, it is the most difficult style of writing to describe in words. The forms of the individual letters are less noticeable and peculiar than in the earlier hands, and approximate to those of the Roman period; and yet the general aspect of the writing is unmistakably Ptolemaic. It lacks the roundness of the Roman style, and the letters are of a uniform thickness, without modulation of broad and thin strokes, with a tendency to thickness and blackness throughout. Of the individual letters, A is generally small, and its loop becomes either a round spot or a straight line; H generally has a ligature attached to its last stroke, whether there is another letter for it to be linked with or not; the same

PLATE III.

is the case with N; Y fluctuates between its earlier shape
and one more resembling a Y, of which the left-hand arm
is generally longer than the right. But, on the whole,
little reliance can be placed in the forms of single letters at
this period; while, on the other hand, the general appearance of this small, even cursive can hardly be mistaken.

The first century was, until quite lately, the most obscure
period in the whole history of papyrus-palaeography; and
it cannot even yet be said to be adequately known. On
the one hand there were a few documents dated between
100 and 80 B.C., which carried on the tradition of the
preceding century without much recognizable variation; and
on the other there were a few which belonged to the last
decade of the century, in which the Roman cast of hand was
already well developed. The interval of transition is now
precariously bridged by some papyri acquired by Messrs.
Hogarth and Grenfell during their campaign on behalf
of the Egypt Exploration Fund in 1895–6. The chief
conclusion to be derived from these is that the transition
was very rapid. In the early part of the century the
Ptolemaic cursive is seen to be breaking up. It becomes
less firm and regular, and loses its sense of style. Under
Ptolemy Lathyrus and Auletes it is an ugly and a broken
hand. In the middle of the century (if the dates assigned
to these documents, which are sometimes uncertain, are to
be trusted), several forms of letters which are characteristic
of the reign of Augustus are found intermixed with
Ptolemaic forms; and by the close of the century the
Ptolemaic forms have entirely disappeared, and the writing
is distinctly Roman. The only criterion that can be given
for the determination of MSS. of this period is to remember
that it is an age of transition, and to look for an intermixture of forms. The styles of the end of the second
century B.C. and of the beginning of the first Christian
century must be known first, and the intermediate stage
will then be recognizable.

The first century B. C.

If one special characteristic is to be assigned as, more
than any other, marking off the Roman period from its
predecessor and its successor, it is *roundness*. Straight,
stiff lines are avoided; curved, flowing strokes take their
place wherever possible. Ligatures, which in the Ptolemaic
cursive hand were generally straight lines, are generally
curved in the Roman hand. There is more modulation
of strokes, and the somewhat thick and black aspect
of Ptolemaic writing disappears. It also becomes usual
to write letters as much as possible without raising the
pen. This is especially noticeable in the case of Є, the
cross-bar of which is now very commonly attached to its
upper curve (C⁻, C⁄); and this formation may be taken
as an almost conclusive proof of a Roman date. In early
Roman documents, on either side of the beginning of the
Christian era, Y is very noticeable for the deep curve
of its upper part and rather prominent development of its
tail (ʮ). Other letters besides Є exemplify the tendency
to form letters in single strokes. The oblique strokes of K
become a curve attached to the bottom of the upright
stroke by an obliquely-rising ligature, so that the whole
letter assumes a shape resembling a written *u*. An almost
identical shape is often assumed by B, which begins in this
period to be frequently formed with open top, though the
capital form continues contemporaneously. The cross-bar
of N becomes a curve uniting the tops of the two upright
strokes (ᴎ). A change is also observable in T, which
becomes fork-shaped in many instances, and sometimes
almost has the form of a V (Υ, ᴠ). Even φ is written
without raising the pen, being formed of a semi-circular
curve, the end of which is attached to the top of the
perpendicular stroke, and the latter not unfrequently falls
outside the curve altogether; at other times the circle
is represented by a stroke resembling an *s* lying on its
side, through which the perpendicular stroke descends.

These characteristics run, more or less, through all the

PLATE IV.

OFFICIAL LETTER—A.D. 15.

PLATE V.

POLL-TAX REGISTER.—A.D. 72-73.

Roman period, serving to differentiate it from the Ptolemaic
and Byzantine ages, but not sufficing for the accurate
dating of MSS. within the first three centuries. This is
a matter partly of general appearance, partly of variations
in a few individual letters. Documents of the reign of
Augustus and his successors as far as the middle of the
first century are generally written in a bolder and more
angular hand than those of a later date[1]. The forms
of the letters are more pronounced, as in the case of Υ,
mentioned above, Ε and Κ. In the middle of the century
these angles have been smoothed away, and the round,
graceful character of the best Roman cursive is fully
developed. Indeed the hand may be said to be at its best
between the years 50 and 100. The specimen here given
(Plate V), from a poll-tax register of A. D. 72-3 (Brit. Mus.
Pap. CCLX), is a good example of the larger hand of the
period, and one which will be found useful in assigning
a date to some literary papyri in a later chapter. At the
same time a smaller hand, and one much more difficult
to read, came into fashion in the reign of Domitian; and
small hands, though not unknown earlier in the Roman
period, become predominant from this point until the end
of the second century or later; increased cursiveness going,
as usual, with reduced size of the letters. An example,
from the extreme end of the first century, is given in
Plate VI (Brit. Mus. Pap. CXLIII).

Two letters are especially useful as indications of manuscripts written in the latter part of the first century and
the first half of the second. These are Η and Ϲ. The
former letter, during this period, has two quite distinct
forms. One is the familiar Η-shape, in which the cross-bar
usually rises slightly and the second upright descends from
it in a shallow curve; this is found more or less throughout
the period, and in itself contains no criterion of age. But

---

[1] Plate IV, an official document from the beginning of the reign of Tiberius
(Brit. Mus. Pap. CCLXXVI), is a fairly characteristic example of this style.

by the side of this there is another form, consisting of a
horizontal curve with an almost perpendicular tail (˥, ˩),
which is characteristic of a comparatively limited space
of time. In appearance it resembles the Y of the same
period, and is often only distinguishable from it by the
fact that its tail is either perpendicular or curls to the left,
while the tail of the Y curls to the right. This form of
H is, in point of origin, a modification of the ordinary
H-shape, in which the first perpendicular stroke is so
much slurred as almost to disappear, and accidental ap-
proximations to it are occasionally found in much earlier
documents; but to these no importance can be attached,
and the real range of its currency may be stated as being
from about A.D. 50 to about 160[1]. There are sporadic
instances later than this, as is inevitably the case with
every palaeographical characteristic; but the material for
the first three centuries is now so plentiful that we are
justified in asserting that the practical disappearance of
this form at about the date named is not accidental,
or a fancy based on the absence of evidence, but really
represents a fact in palaeography.

In the case of C the data are less precise, but still useful.
The C of the Ptolemaic period is upright and has invariably a horizontal top (⊂). In the early Roman period the
upright form is still maintained, and often the horizontal
top as well, though the Roman fondness for curves shows
itself in the increasing use of a rounded top. Towards
the end of the century the curve has a tendency to be
carried further over, and the letter assumes a tumble-down
appearance (⌒, ⌒, ⌒); while in the next century it sometimes
even passes the horizontal stage, and becomes a backward-

---

[1] It is found very plentifully in the first and fourth hands of the
Aristotle papyrus (indeed it was here that its existence was first noted),
written about A. D. 100, and also in the accounts on the *recto* of the same
papyrus, which were written in A. D. 78-9. Plate VI, a receipt of A. D. 97,
not only has this letter but generally resembles the style of the Aristotle
MS., and confirms the date assigned to it.

PLATE VI.

RECEIPT.—A.D. 97.

PLATE VII.

RECEIPT.—A.D. 166.

facing curve ($\cap$, $\supset$; e.g. $\circlearrowleft$ = $\alpha\varsigma$). The upright C with flattened top continues simultaneously with this form throughout the whole period, and consequently is not to be taken as a sign of early date; but the tumble-down C, according to the fairly plentiful evidence which we now have on the subject, can hardly be earlier than the end of the first century. Early in the third century the upright form seems to have been universally resumed, though in a larger and coarser type.

For the second century palaeographical material is more plentiful than for any other period in the whole history of writing upon papyrus, not a single year being unrepresented by at least one accurately dated document. It is, of course, useless to pretend to lay down precise laws for the discernment of documents of successive decades, but the general lines of development may be indicated. In papyri of the reigns of Trajan and Hadrian, the letters are usually of fair breadth, and are reasonably well formed; while in the reign of Antoninus Pius, and still more of Marcus Aurelius, the increased cursiveness of the prevalent hand leads to the compression of letters and often to their slovenly formation. A very small and very cursive hand is especially characteristic of the reign of Marcus [1]; and in many of the receipts and leases of this period it requires little less than divination to make out the scribe's intention—especially if, as often happens, illiteracy be added to cursiveness. A reaction from this very minute style is manifest in the reign of Commodus and under the emperors of the first part of the third century; but though the writing gets larger, there is no regeneration in the formation of the letters. On the contrary, from this point may be dated the break-up of the Roman hand. The writing becomes rough and coarse, and often extremely ugly. Letters are less formed and worse formed, and the writing straggles unevenly over the page.

Second century.

Third century.

[1] See Plate VII, a receipt of A. D. 166 (Brit. Mus. Pap. ccxxxii).

To this rule, as to all others, there are exceptions; one
(Brit. Mus. Pap. cccxxii) so marked as to make us dis-
trust the cogency of the argument which, on the ground
of the occurrence in it of the name of Aurelius, assigns it
to this period [1].

About the second quarter of the third century a marked
improvement takes place, for which no adequate reason can
be assigned, unless it be a mere accident in the survival
of evidence. As a rule, the variations in handwriting in
Egypt curiously correspond to the changes of government.
The rise of a new government is accompanied by the appearance of a new style of writing, and the decay of the
writing goes hand in hand with that of the administration.
The rise of the Ptolemies, the collapse under Auletes and
his fleeting successors, the advent of the Romans and
the firm rule of the early emperors, the decline when
Commodus and Caracalla succeeded Antoninus and
Marcus, the reorganization of the empire under Diocletian, all have their palaeographical parallels in the documents of the period; but there seems to be no reason
why the chaotic years about the middle of the third
century should show any improvement on their prede-
cessors. The fact, however, remains visible in such
evidence as we possess, and there are several extant
documents between A.D. 250 and 280 which are written
in tidy and even ornamental hands (*e.g.* Brit. Mus. Pap.
ccclI, Paris Pap. 69 c, and a Berlin papyrus shown in
plate xiv of Wilcken's *Tafeln*).

At this point in the history (about the middle of the
third century), the evidence which has been so plentiful
since the beginning of the Christian era suddenly becomes
extremely meagre. Indeed for the important period of
transition in the reign of Diocletian evidence was wholly

---

[1] The use of the name Aurelius by private individuals was granted in
A. D. 212, and thenceforward it becomes so common as almost to be universal; but there are certain instances of its use as far back as A. D. 175.

wanting until the acquisition by Mr. Grenfell, in 1895-6, of a small group of documents (now in the British Museum) from the years on either side of the turn of the century. For the present, therefore, one can only speak with some uncertainty as to the details of the change. That a change, however, did take place, is manifest on a comparison of papyri of the middle of the fourth century with those of a hundred years earlier, the difference between them being quite as marked as that between papyri of the reigns of Auletes and Augustus. The transition period is brief, but has a well-marked character of its own. Writing becomes smaller again, though not so small as in the second century. It is compressed laterally, and the letters are stiffer and more angular in formation, with a tendency to lean forward. It looks as if the formalism of the Byzantine age began to impress itself upon its scribes even from the very beginning of the revolution brought about by Diocletian.

For the fourth century nearly all the evidence at present extant consists of a single group of papers, containing the correspondence of a certain Abinnaeus, who was the commander of a body of auxiliary cavalry quartered at Dionysias, in the Fayyum, for some years on either side of A.D. 350. The larger part of this collection is now in the British Museum [1]; the rest are at Geneva. To these have to be added two examples of the more ornamental writing of the period, published in Wilcken's *Tafeln* (Nos. xv. and xvi.), and a few magical and other undated papyri which, in the light of the Abinnaeus papyri, we can now confidently assign to this century. There are also some good specimens in the Rainer collection at Vienna, but these are not yet published; and the same is the case (so far as facsimiles are concerned) with the fourth-century documents among the Oxyrhynchus

Fourth century.

---

[1] Many of these are reproduced in the atlas of facsimiles accompanying vol. ii. of *Greek Papyri in the British Museum*; and one, which is a good example of the better hands in the collection, is given in Plate VIII.

papyri and at Berlin. The most striking general characteristic of fourth-century writing is an increase of size and
(in the better written and representative examples) of calli-
graphic style and ornament. The letters are boldly formed,
with sharp angles and prominent strokes. The better hands
are generally composed of fine and delicate lines, which
emphasize the sharpness of the angles and the boldness
of the curves; but the more illiterate hands, of which
there are many among the Abinnaeus papyri, are often
very thick and coarse. Of single letters the most noticeable are K and O, on account of their increased size and
prominence. The sharp angles and projecting upper
strokes of K almost always catch the eye in documents
of this period; while O, which in Ptolemaic and Roman
writings is often reduced almost to a dot, is now a large
and conspicuous letter. The right-hand stroke of Δ is
often widely separated from the rest of the letter; and
sometimes a d-shaped letter takes the place of the familiar
form. The upper half of ε is a straight oblique stroke,
pointing to the right (ϵ); while η has already acquired
the h-shape which belongs to it throughout the Byzantine
period. C once more stands upright, generally with a
flattened top, as in the Ptolemaic and early Roman age.
Finally Y is v-shaped, and ceases to be a conspicuous
letter. The general impression left by a page of good
writing of this period is one of regularity and order,
though not so mechanical as it subsequently became.

III. By-
zantine
Period.
Fifth
century.

The largest unexplored tract now left in the history
of cursive writing on papyrus is that from about A.D.
360 to about A.D. 500. A few documents from the
Great Oasis, recently acquired by the British Museum from
Mr. Grenfell, represent the years on either side of 400, and
tend to show that the style current in the middle of the
century had not greatly changed by then. The Rainer collection possesses some fragments of the fifth century, but
these have not been published, and are not available for

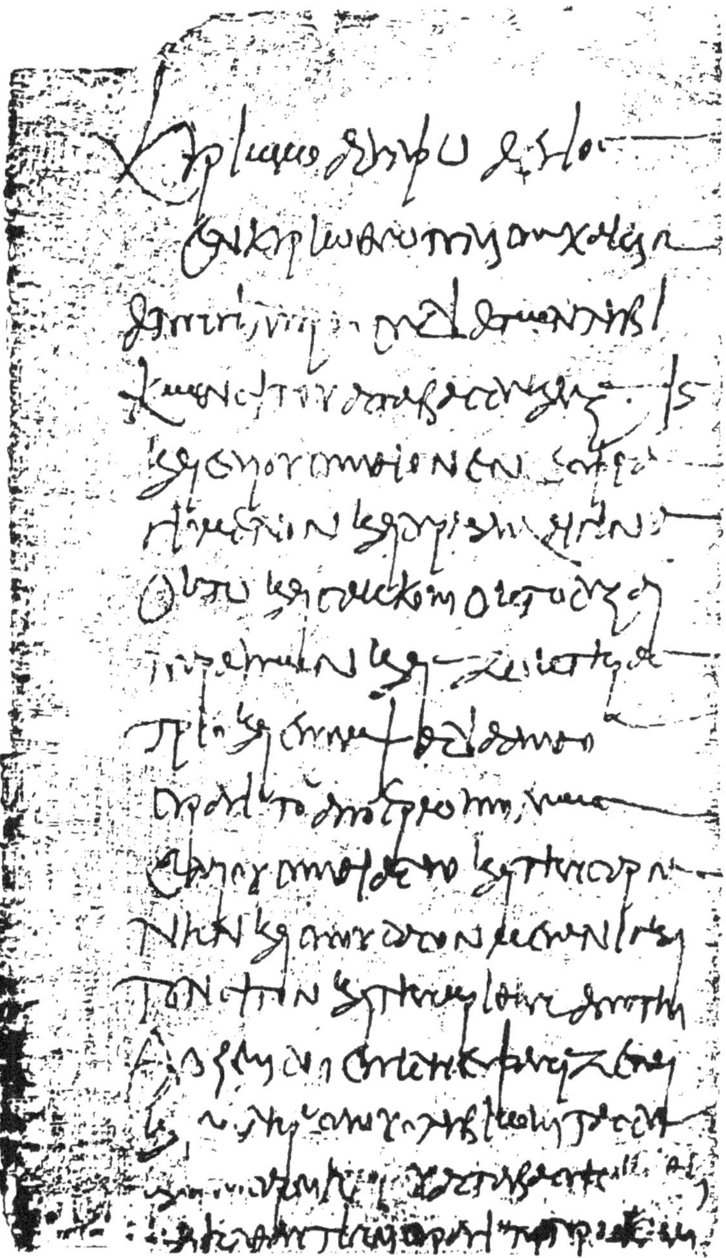

LETTER.—*circ.* A.D. 350.

comparison. It is not until close to the end of the century, in A. D. 487 and 498, that dated materials are again accessible, and these cannot be separated from the sixth century in general. There, on the contrary, material is once more plentiful. The first Fayyum find, which has supplied documents in thousands to Vienna and Berlin, and in lesser quantities to Paris, Oxford, and London, consisted almost entirely of papyri belonging to this later Byzantine period, extending from the beginning of the sixth century to the end of Greek writing in Egypt. Unfortunately, the number of precisely dated documents of this period bears a very small proportion to that of the undated or insufficiently dated[1]. This is largely due to the practice of dating by the fifteen-year period known as the indiction (see below, p. 54), which by itself is quite useless for the purpose of determining the age of a document after the lapse of a few years. The consequence is that, while it is generally very easy to assign a hand to the later Byzantine period, it is very difficult to determine whether it is of the sixth or seventh century. The difficulty is increased by the conservatism of Byzantine scribes, who adhere for a considerable time to the same type of hand.

Speaking generally, the fully-formed Byzantine hand is a large, well marked, and rather handsome hand; not so delicate as the best examples of the fourth century, but regular, with ornamental strokes and curves, and with an unmistakable air of formality. In the best specimens the writing is upright and square, with plenty of width, so that any given piece of Byzantine papyrus contains much less writing than a similar piece of the Roman, or even of the Ptolemaic period. Many of the letters, also, are formed in quite peculiar manners. The

Sixth and seventh centuries.

---

[1] This defect will be rectified when dated facsimiles of the Oxyrhynchus papyri of the Egypt Exploration Fund are published. The specimen shown in Plate IX is not dated, but seems from its style to belong to the sixth century.

loop of α is large and generally open at the top, resembling a modern α more than any α of the earlier periods. β is generally a long, irregular oval, open at the top, and with a short tail projecting downwards from the bottom right-hand corner. The minuscule form of δ, resembling a Latin d, is increasingly common, and has a very long upright stroke; but the uncial form is also found. The upper half of ε is still more pronounced than in the fourth century, projecting obliquely for a considerable distance above the line (є). η is generally h-shaped, with a very long upright stroke, like δ. Similarly, the first stroke of κ projects far above the line, the rest of the letter being u-shaped. The left-hand stroke of λ comes far below the line, and is often widely separated from the right-hand stroke. μ is in its minuscule form, much as it is here printed, with its first stroke perpendicular and stretching far below the line. ν varies between the uncial type, in which case the junction of the oblique stroke with the second upright often takes the form of a curve (Ц, Ш), and a cursive form resembling a Latin n. σ is well rounded, generally divided in the middle, with the upper half sometimes rather exaggerated. τ, if not of the ordinary shape, has a long tail and is deeply forked at the top. υ is small and v-shaped, and not unfrequently becomes little more than a curved ligature in combination with other letters. The upright stroke of φ is generally united to the circle by a well-rounded curve, and the whole letter is usually large and prominent, —as, it may be remarked, it likewise is in vellum uncials of the same period.

Besides this upright hand, a sloping style is also found in use during these centuries. A papyrus in the British Museum (CXIII. 5 b) shows it in existence in A.D. 542, but it is especially characteristic of the later years of this century and of the seventh century. The shapes of the individual letters are much the same as those just described,

PLATE IX.

LEASE.—6TII CENT.

but they assume a marked slope to the right, become smaller and less square, and are laterally compressed. α is more closely connected with the letter which follows it, and is often a mere loop, like the υ described above. The upper half of ϵ is more rounded, and, with the central cross-stroke, is often separated from the lower half. The left-hand stroke of λ is still more prolonged, contributing much (like the up-strokes of δ and η) to the sloping appearance of the writing. The tail of the first stroke of μ is shortened. υ is not unfrequently written above the line, in the shape either of a very wide and shallow curve, or sometimes of a straight line.

This sloping hand is especially characteristic of the later part of the Byzantine period, but it did not extinguish the upright style of writing, which is found even in the seventh century. For instance, the British Museum has a good example of it dated in the year 619 (Pap. ccx). Hence it is unusually difficult to assign a precise date to any undated document of this upright type, and one must commonly be content to define it merely as 'sixth or seventh century'; though in the later examples the characters are generally less firm and precise than in the earlier.

The two types of hand which have just been described are, both of them, large, and cover a good deal of ground; but there is also a small type of hand, which is used chiefly for accounts and receipts. These are usually assigned to the seventh, and sometimes to the eighth, century; but their age is really very uncertain, since few are dated in any other way than by indictions. The forms of the letters are much the same as those of the larger hands, and the projecting strokes of such letters as δ, η, λ, μ, are equally marked; but their smaller size calls attention to the fact that the Byzantine hand, at any rate in its later stages, is definitely a minuscule hand. Whether written large or small, the forms of

nearly all the test-letters are now minuscule; and the
special interest of these small hands is that they indicate
the way to the transition from the minuscules of the
papyri to those of the vellum MSS. of the ninth and
tenth centuries. More will have to be said on this point
in a later chapter; for the present it must suffice to state
the proposition that, while the literary hand of the Roman
period is the parent of the vellum uncial, the Byzantine
non-literary hand is the parent of the vellum minuscule,
which comes to the front in the ninth century as a book-
hand for the first time, and establishes its supremacy
in the tenth.

The end of Greek writing on papyrus.

The exact end of Greek writing upon papyrus cannot
be fixed; but it is certain that the Arab conquest of
Egypt in A.D. 640 gave it its death-blow. Documents
that can be placed with certainty later than this date
are rare; probably there are more at Vienna than
anywhere else, but they are not yet accessible. The
latest papyrus of any length with a precise date is of the
year 683; but one long document in the British Museum
(Pap. LXXVII) is probably of still later date than this[1]. This
is the will of Abraham, bishop of Hermonthis and head
of the monastery of St. Phoebammon at Djemé, near
Thebes. It forms one of a group of documents, the rest
of which are written in Coptic; one of these is dated
in the year 786, and several others are shown by internal
evidence to belong to about the same date. The will
of Abraham appears to come near the beginning of the
series, but there is nothing to show that it falls outside
the eighth century. It is a large and fairly upright
hand, showing that the traditions of the sixth century
lasted on even till this late date; only the somewhat
broken and degenerate look of the writing, the roundness
and looseness of the shapes of the letters, distinguish

[1] Complete facsimile in atlas accompanying vol. i. of *Greek Papyri in the
British Museum*.

it from the square and precise appearance of the hand
from which it is descended. It marks, however, the close
of Greek writing in Egypt; and it is noticeable that the
bishop, whose testament it is, is expressly said in it to
be ignorant of the Greek language. The Greek language
was, in fact, gradually extinguished by the Arab conquest:
and with it disappears our knowledge of Greek writing
on papyrus, since in no other land than Egypt has the
brittle material survived to our own time. The full
history of the transition from papyrus to vellum can
never be written, for want of the materials.

One detail of importance in connexion with non-literary   The dating
papyri must be mentioned before closing this sketch of   of papyri.
their history. This is the manner in which they are
dated. The formulas of dating differ, like the hands
themselves, in the Ptolemaic, Roman, and Byzantine
periods. In a Ptolemaic document the date is given by
the regnal year of the reigning sovereign, and the full
formula for this contains not only the name of the king
himself, but also a list of the priesthoods of all the defunct
Ptolemies; for example (Brit. Mus. Pap. DCXXIII).

Βασιλευόντων Κλεοπάτρας καὶ βασιλέως Πτολεμαίου Θεῶν
Φιλομητόρων Σωτήρων ἔτους ῆ, ἐφ' ἱερέως τοῦ ὄντος ἐν 'Αλεξανδρείᾳ 'Αλεξάνδρου καὶ θεῶν Σωτήρων καὶ θεῶν 'Αδελφῶν
καὶ θεῶν Εὐεργετῶν καὶ θεῶν Φιλοπατόρων καὶ θεῶν 'Επιφανῶν
καὶ θεοῦ Εὐπάτορος καὶ θεοῦ Φιλομήτορος καὶ θεοῦ Φιλοπάτορος
νέου καὶ θεοῦ Εὐεργέτου καὶ θεῶν Φιλομητόρων Σωτήρων,
ἱερουπώλου Ἴσιδος μεγάλης μητρὸς θεῶν, ἀθλοφόρου Βερενίκης
Εὐεργέτιδος, κανηφόρου 'Αρσινόης Φιλαδέλφου, ἱερείας 'Αρσινόης
Φιλοπάτορος, τῶν οὐσῶν ἐν 'Αλεξανδρείᾳ, ἐν δὲ Πτολεμαΐδι τῆς
Θηβαΐδος ἐφ' ἱερέων καὶ ἱερειῶν καὶ κανηφόρου τῶν ὄντων καὶ
οὐσῶν, μηνὸς Μεχεὶρ ῑᾱ ἐν Κροκοδίλων πόλει τοῦ Παθυρίτου.
All this merely describes the year which we indicate as
B.C. 109. But since such a formula might be accused
of being cumbrous, and of including a good deal of un-

important matter, it is very often reduced to the simple statement ἔτους ῆ, Μεχεὶρ ῑα : which errs as much from brevity as the other does from length, since it omits the name of the king, and it is not always that the reign can be identified with certainty.

Roman dates are more business-like, but they too have a longer and a shorter formula. Dates are given by the regnal year of the emperor; but in both Ptolemaic and Roman dates it must be remembered that the year always begins with the 1st Thoth (= 29th August). Thus the first year of a sovereign lasted only from his accession to the 1st Thoth next ensuing. The longer Roman formulas are of the type ἔτους ῆ αὐτοκράτορος καίσαρος Τίτου Αἰλίου Ἁδριανοῦ Ἀντωνίνου Εὐσεβοῦς Σεβαστοῦ, Μεχεὶρ ε̄, the shorter ἔτους ῆ Ἀντωνίνου Καίσαρος τοῦ κυρίου. In neither case can there be any doubt as to the precise date intended, and it is only under some of the later emperors that the accumulation of titles (Γερμανικοῦ Παρθικοῦ Δακικοῦ Βρεταννικοῦ Μεγίστου, and the like) becomes cumbersome.

This useful and practical system of dating was, however, abandoned at the time of the revolution under Diocletian. In the first instance dating by the consuls of the year was substituted; but in the year 312 the system of the indiction was instituted. This was a fifteen-year period, beginning on different days in different parts of the empire. At Constantinople it began on the 1st September, in Egypt on a fluctuating date about the middle of June, at the time of the rising of the Nile [1]. Its origin is uncertain, but it is probably a modification of the fourteen-year census-period in use under the Romans. The formula of dating by the indiction is Φαρμοῦθι ζ̄, ῑε ἰνδικτιῶνος, meaning the fifteenth year of the current indiction-period; but as the indiction-periods are not themselves numbered,

---

[1] See *Greek Papyri in the British Museum*, i. 196-8.

this dating is absolutely useless from our point of view. Sometimes the names of the consuls or the regnal year of the emperor are added, but this is not by any means the regular practice; and consequently the dating of Byzantine documents remains an obscure subject, in spite of the immense mass of material for the period which is in existence. The publication of select dated facsimiles from the Vienna and Berlin collections is much to be desired, in order to remove this obscurity; but perhaps a speedier revelation may be looked for from the Egypt Exploration Fund.

CHAPTER IV

In passing to the consideration of literary papyri, one enters what is at once the most important and the least familiar part of the subject. Non-literary papyri, as will have been seen from the preceding chapter, are plentiful and fairly well supplied with precise dates; literary papyri are comparatively rare and can seldom be dated, even approximately, on other than purely palaeographical evidence. Nor can all manuscripts which contain literary works come properly into consideration here, since some of them are not written in formal book-hand at all, but in the ordinary private or non-literary hand of the day. Such is, notably, the papyrus containing Aristotle's *Constitution of Athens*; also a large medical manuscript in the British Museum, containing considerable extracts from the Ἰατρικὴ Συναγωγή of Menon, the pupil of Aristotle, sometimes ascribed to Aristotle himself; and an astronomical treatise derived from Eudoxus in the Louvre, of the second century B.C. In this way the bulk of the materials at our disposal is reduced; but on the other hand there are a few documents of a private or business character which are written by professional scribes in literary hands; and these, bearing, as they do, precise dates, are very valuable items of evidence for the construction of the history of papyrus-palaeography on its literary side.

Indeed the document to which is generally assigned

the first place in any treatment of the subject is itself, to some extent, one of this description, being a non-literary composition written in an uncial hand. This is the so-called 'Curse of Artemisia,' consisting of the imprecations of a woman against the father of her child, who has deserted her, written in an extremely archaic uncial hand. The papyrus was found on the site of the Serapeum at Memphis (though not with the papers of the recluse Ptolemy, mentioned in the last chapter), and is now in the Imperial Library at Vienna[1]. It has been assigned by Blass and Wessely to the fourth century B. C., by Thompson to the early part of the third; and there is no doubt that the forms of the letters, recalling as they do those of inscriptions upon stone, have a very early appearance. The letters are stiff and angular, with few curves; A and E are of the square epigraphic shape, the latter generally having the top horizontal stroke very long; Θ is a circle enclosing a dot (⊙); O is small; C is of a form intermediate between the Σ of the inscriptions and the C of the papyri (C, ⟨); ω retains much of the epigraphic form (ᴧ, ᴧ); a colon is used for purposes of punctuation, as in inscriptions and a few early papyri. What militates against the value of this document for palaeographical purposes is its extremely rough and untrained appearance. It is not the work of a professional scribe, but the writing of an uneducated woman who uses uncial letters because she can form no others. If Greek writing were at this period just issuing from the purely epigraphic stage, stress might fairly be laid upon the use of epigraphic forms as a proof of age; but since we now know that men wrote freely and easily upon papyrus long before this time, this argument falls to the ground. Artemisia used letters like those employed in inscriptions for the same reason that an illiterate person always uses capitals, because such

The Curse of Artemisia.

[1] Facsimile in the Palaeographical Society's publications, vol. ii. pl. 141.

letters were commonly before her eyes in public places, while she had probably seldom seen a book. It is on evidence of a different character that the early date of this document can alone be maintained. The traces of Ionic dialect ('Αρτεμισίη, ἰκετηρίη), the assimilation of the final consonants of prepositions to the first letter of the following word (ἐμ ποσεράπι, ἐγ γῆι), and the occurrence of forms of letters which are found not only in inscriptions but in papyri known to belong to the third century, such as those of Є, O, ⲱ, described above, may be admitted as legitimate evidence that this is a genuinely early document; but it is not one on which we could base any sound argument as to the character of contemporary MSS., if we had no knowledge on this point from other sources of information.

The third century B. C. The Petrie papyri.   Our knowledge of the literary palaeography of the third century B. C. is based, in fact, entirely on the papyri discovered by Mr. Flinders Petrie in the mummy-cases of Gurob, to which reference has already been made more than once. Among them, in addition to the non-literary documents described in the last chapter, were several fragments of literary works; and it is a fair conclusion that these are of about the same age as the non-literary papers among which they were found. The argument is, of course, not quite decisive. The papyri out of which the Gurob cartonnages were made were, no doubt, waste paper at the time when they were so used; and it may be questioned whether literary and non-literary documents found in the same waste-paper basket are likely to be of the same age. On the one hand it has been argued that a well-written MS. of Plato or Euripides would not be thrown away nearly so soon as mere business papers of ephemeral interest, and hence that the fragments here in question may easily be as early as the fourth century. Against this it has been contended that such documents as wills, leases, and receipts, which form the titles to property, would probably be preserved for several generations, while a copy of a literary work might be thrown away at any time if it were found defective or injured in any way. These arguments may perhaps be allowed to neutralize one another. Titles to property do not seem to have been preserved with the same tenacity as in modern days, and were indeed unnecessary when a register of ownership was kept by the government; while, if some of the literary MSS. were of a period much earlier than that at which they were converted into mummy-cases, we should expect to find a perceptible gradation of hands in the various fragments, some being old while others were more recent. This is not the case to any important extent; and we are therefore justified in assigning the literary fragments among the Petrie papyri to the same period as their non-literary companions, namely to the third century.

To the Petrie fragments may now be added a few scraps acquired by Mr. Grenfell in 1895-6, which, though bearing no independent proof of their date, evidently belong to the same period[1]. What we have, therefore, in all, as our material for judging of the literary palaeography of the third century, consists of the following: four columns of the *Antiope* of Euripides[2]; about twelve columns, besides minor fragments, of the *Phaedo* of Plato[3]; five columns of the *Laches* of Plato[4]; and about a score of minute fragments of various works[5], all of them unknown with the exception of a few pieces of Homer. Some of these, however, can hardly be reckoned as literary hands. With these, and with the evidence derivable from the non-literary

---

[1] Facs. in Grenfell, *Greek Papyri*, vol. ii. pl. i; now in British Museum.

[2] Facs. in Mahaffy, *Flinders Petrie Papyri*, part I. pll. i, ii; now in British Museum.

[3] Facs. *ib.*, pll. v–viii; partial facs. in Plate X and in Pal. Soc. vol. ii. 161; now in British Museum.

[4] Facs., Mahaffy, part II. pll. xvii, xviii; now in Bodleian Library.

[5] Facs., *ib.*, part I. pll. iii, iv, ix, x, xxv; part II. pl. xvi; and Grenfell, l. c.; now divided between the British Museum and the Bodleian.

papyri of the same period, the history of writing during the century has to be reconstructed.

An examination of these texts shows, in the first place, that they fall into two groups, one consisting of very small hands, while the others are larger and present generally a rougher appearance, apart from differences in the formation of individual letters. The more important, and also the best-written, texts are all in the smaller hand, which may fairly be taken as representing the writing in current use for books intended for public circulation. They include the *Phaedo* and *Antiope* MSS., a few of the Petric fragments, and all those of Mr. Grenfell. The order of antiquity among them can only be determined by the more or less archaic appearance of the writing. In the beginnings of writing on papyrus it cannot be doubted that, in formally and carefully written MSS., the shapes of the letters were nearly identical with those in contemporaneous use in inscriptions; and the greater or less occurrence of epigraphic forms, in a MS. written, not by an uneducated person (as in the case of the Artemisia papyrus), but by a trained scribe, may be taken as evidence for a relatively earlier or later date. Applying this test, the earliest examples among the papyri now extant would seem to be (as Mr. Mahaffy and Mr. Grenfell have already noticed) a fragment of a prose narrative of the adventures of Heracles, among the Petric papyri [1], and two minute scraps among Mr. Grenfell's. The most notably early among the forms of letters in these fragments are the square E with broad top; Θ with a central dot instead of a cross-stroke; ω in its transition forms, between Ω and ω; and especially Ξ, formed of three parallel horizontal bars with a perpen-

Early fragments.

[1] An alphabet from this fragment is given by Mahaffy in his *Flinders Petrie Papyri*, part I, p. 65; and an alphabet from the largest of Mr. Grenfell's fragments forms col. 1 of the table given in Appendix I to this essay, to which the reader may be referred for illustrations of the descriptions of the individual letters which follow.

dicular stroke cutting them at right angles, and C in the
angular epigraphic shape. These last two characters belong
only to the Grenfell fragments; the Heracles fragment has
no example of a Ξ, and its C is semi-circular. Of the other
letters it is sufficient to observe that A is of the uncial
shape, like a modern capital A ; H is rather rounded
on both sides; the two strokes composing Λ meet at the
top; M is somewhat deeply curved in the Grenfell fragment
and angular in the Heracles: O is very small; Π is rather
broad, with the right leg inclined to be shorter than
the other; and Y is shaped as here printed. Not all these
letters are of value for determining the status of these
texts among other third-century, or even among Ptolemaic,
papyri; but it is necessary to note them for comparison
with the forms which came into existence in later periods.

Somewhat later than these fragments—perhaps about
the middle of the third century—must be placed the two
great treasures of Mr. Petrie's find, the *Phaedo* and the
Antiope. In these cases we have some substantial part
of each MS. preserved, and are in a better position to
judge of their general appearance. The hand is not the
same in both, that of the *Phaedo* being noticeably better
and more careful; but in size and formation of letters
they are alike. The writing is extremely small, with
the object, no doubt, of making neat and handy volumes;
the complete *Antiope*, supposing it to have been a play
of between 1400 and 1500 lines, would have occupied a roll
of about 12 feet in length. Of the individual letters[1],
A is of the uncial shape, in the *Antiope* sometimes (but not
always) approaching nearer to the epigraphic shape by
having the cross-bar bent downwards into an angle. The
Plato also sometimes has the square E with elongated top;
but elsewhere in the Plato, and always in the Euripides,
the rounded form of the letter is found. Z (which does

The Phaedo and Antiope MSS.

---

[1] See alphabets in columns 2 and 3 of the table in Appendix I.

not occur in the earlier fragments described in the last paragraph) has the form, peculiar to the third century, of two parallel bars joined by a perpendicular stroke; Θ has a cross-stroke, though it does not always accurately fit the circle; Λ and M are of the forms already described; Ξ is composed of three parallel and unconnected strokes, and this, it will be found, is characteristic of the Ptolemaic age alone, and forms a useful criterion for distinguishing Ptolemaic MSS. Of the remaining letters, Π is very broad in the *Phaedo* (it will be remembered that this is a feature of third-century non-literary papyri), less so in the *Antiope*; C is small and rounded, the top sometimes slightly flattened; Ѡ is of the minuscule shape, the second loop being often incompletely formed.

The beauty of these hands, and especially of the Plato, has been eulogized in terms which are, perhaps, somewhat exaggerated; not unnaturally, since, in the state of our knowledge at the time of their discovery, it was a surprise to find writing of so early a date which showed such freedom of style, combined with so much orderliness and precision. Both Mr. Mahaffy and Sir E. Thompson speak of the Plato, in particular, as a MS. of extreme beauty. Now that the first joy of discovery has worn off, it seems evident that they are not the equals of the best specimens of the Roman period, or even, it may be, of the later Ptolemaic age. There is a general appearance of neatness about these small, yet firm, hands; but there are considerable inequalities in size among the letters, which would be more noticeable if they were written on a larger scale, and they lack the handsomeness of the larger hands which came into favour later. The *Antiope* MS., indeed, hardly achieves the praise of neatness, being decidedly rough and irregular in places, besides being more compressed than the Plato. It cannot be supposed that this is a fair example of the volumes preserved in the great Alexandrian library, though it may represent the

PLATE X.

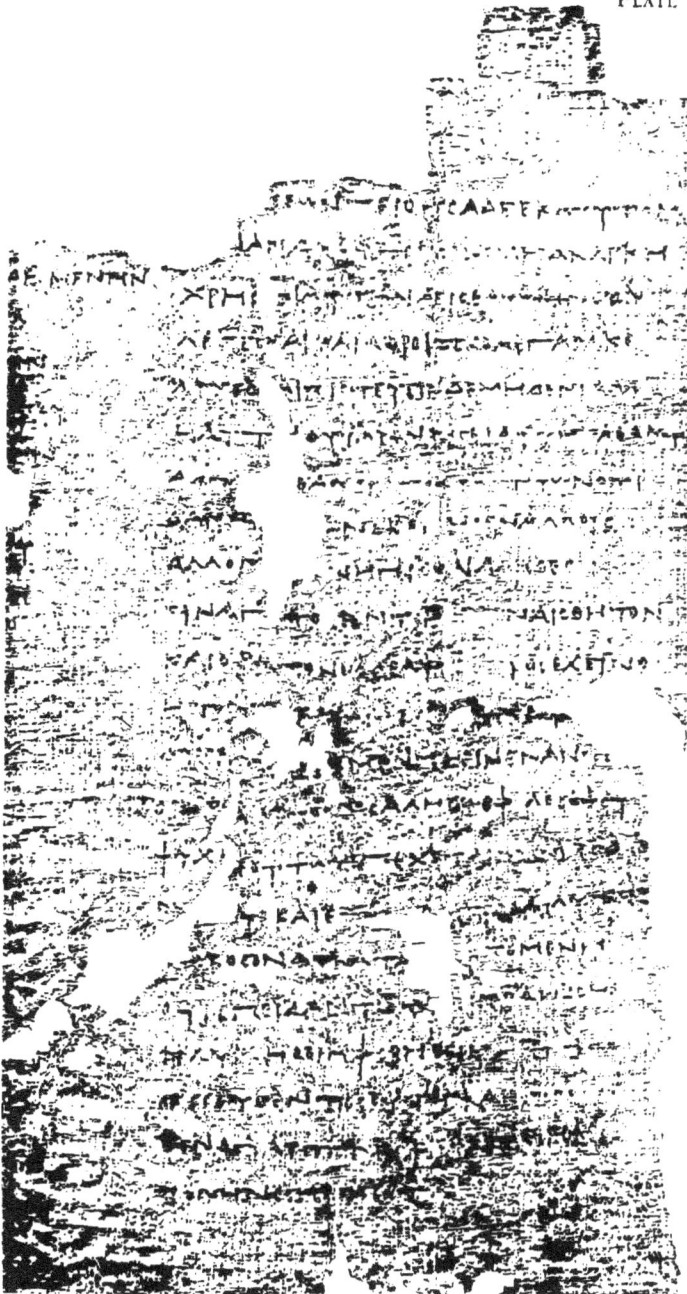

PLATO, PHAEDO.— 3RD CENT. B.C.

style which would pass muster in the book-shops. Nor
do the texts come up to the Alexandrian standard of
accuracy. How far they fall short is a question to which
the answer depends on the solution of the critical problem
as to the authenticity of certain readings which depart
from the vulgate; but apart from these there are more
obvious blunders than can have been admitted in the great
libraries which handed down, in most cases, such pure
texts to the vellum MSS. of the tenth and later centuries, on which our present knowledge of the classical
authors is based. Yet, with all these deductions, the
Phaedo and *Antiope* papyri have a great attractiveness
of appearance, and rank high among papyrus MSS.
of any age; nor would any one deny their extreme value
for the history of Greek writing. With their aid we can
not only realize how Greek books were written in the days
of the early Ptolemies, but can, by a legitimate use of
inference and imagination, picture to ourselves the contents
of the book-shops of Athens in the times of Menander and
Demosthenes, perhaps even of Aristophanes and Sophocles.

Between the literary hands which have hitherto been
described, and the non-literary hands of the same period,
there is no very marked resemblance. Some of the letters,
such as Π and Ξ, and to a less extent M, T, and Y, would
be recognized as Ptolemaic in style, but it would be
very difficult to assign these MSS., on palaeographical
grounds alone, to the same period as the non-literary
documents which bear dates in the reigns of the early
Ptolemies. There is, however, another type of hand to
which this statement applies much less fully. Several
of the literary MSS. among the Petrie papyri are
in larger hands, approximating to those of the contemporary non-literary documents. Chief among these
are the fragments of the *Laches* of Plato, now at Oxford,
the curious scrap of Homer containing several additional
verses, the fragments of the *Mouseion* of Alcidamas and

The *Laches* MS., and others.

another rhetorical treatise [1]. The other classical fragments published by Prof. Mahaffy likewise belong to this category, but they are so small, and in some cases approach so closely to the non-literary hand, that they can hardly be taken into account. In this class of hands, while the uncial forms of some letters, such as A, H, K, N, are preserved, and the straggling appearance of the non-literary hands is avoided, the writing on the whole is recognizably akin to the non-literary type. The letters are larger and squarer than in the *Phaedo* and *Antiope*; Є, ω, and in some cases M, are definitely of the minuscule form: and the writing is without the style and attractiveness —due. no doubt, to greater care—which characterize the smaller hands described above. The MSS. of this larger type cannot be regarded as having been intended for sale or general circulation. They are rather the work of ordinary scribes, employing a somewhat modified form of the current handwriting of the day, which they have adapted for literary purposes by reducing its excessive breadth and cursiveness, making it squarer and firmer, without, however, achieving much of beauty or regularity.

The second century B. C.   For the second century B.C. materials are, at present, very scarce. There has been no great discovery of documents of this period, comparable to the Petrie 'find'; and the principal groups of non-literary material which have come to light—the Serapeum papyri for the first half of the century and Messrs. Hogarth and Grenfell's papyri for the second half—contribute between them only one MS. which can in any way be classed as literary. In addition to this single MS. there is also one isolated discovery belonging to this period, which is of considerable interest. Both MSS. are now in the Louvre, the last-mentioned, a recent acquisition, being the famous oration of Hyperides against Athenogenes—one of the orator's

---

[1] *Petrie Papyri*, part I, nos. 10, 25.

masterpieces—while the other, acquired at a much earlier date, is a dialectical treatise containing quotations from a number of classical authors, such as Sappho, Alcman, Ibycus, Homer, and Euripides. Both MSS. are shown to be not later than the second century B.C. by the existence of writing on their backs belonging to the Ptolemaic period: while neither of them shows any such marked resemblance to the Petrie papyri as would justify the assignment of them to the third century. The dialectical treatise must, however, belong to the first half of the second century at latest, since the writing on its *verso* is dated in the year 160 B.C. For the Hyperides the only evidence we have to go upon is the statement of Prof. E. Revillout, its first editor, that it bears on its *verso* accounts in demotic, belonging to the Ptolemaic period. No more precise indication of their date has been given; but this is sufficient to show that the text of the oration can hardly be later than the second century, while certain resemblances between its writing and that of MSS. belonging to the following century seem to make it probable that it belongs to the second half of the century rather than the first. From only two MSS., and these exhibiting very different types of writing, it is obviously dangerous to draw any very sweeping conclusions as to the literary palaeography of the period; but they serve to bridge over the gulf between two periods for which our information is more complete.

The dialectical fragment [1] is written in narrow columns of about two inches in breadth, leaning (as in several other early papyri) strongly to the right. Divisions in the sense are marked by slight spaces in the writing, and by *paragraphi* below the beginnings of the lines in which the pauses occur. The hand is small, though not

The Paris Dialectical Fragment.

---

[1] Complete facsimile (not photographic) in the Album to *Notices et Extraits*, plate xi; photographic facsimile of three columns in *Pal. Soc.* ii. 180, and of one column in Plate XI. Alphabet in Appendix I, col. 4.

F

so minute as those of the *Phaedo* and *Antiope*; it is evidently to be regarded as the lineal successor of these hands. The most noticeable modification in general form is a tendency to roundness and curved forms of letters. In this respect a transitional form may be found in a small fragment discovered and published by Mr. Grenfell [1]. The letters are also more even and uniform in size; and in many of their forms they approximate to some of the MSS. found at Herculaneum, which must be at least a century later. Of the individual letters, Λ is the most noticeable, as giving the first example of a form found in some other early papyri. It is a modification of the uncial form, the original left-hand stroke and the cross-bar being formed together into the shape of a sort of wedge (∠), across the open end of which the right-hand stroke is drawn, generally in a curve (ᴧ). It is an intermediate form between the uncial (A) and the fully developed minuscule (ᶛ), which can be written without raising the pen. Another letter which shows a transitional form is Z, in which the middle stroke is neither perpendicular as in the Petrie MSS., nor drawn obliquely from end to end of the horizontal strokes as in later papyri, but is intermediate between these two forms. H and K are strongly curved. Λ is noticeable for its right-hand stroke in some cases (but not in all) being carried beyond the point of junction; again a sign of transition, since this is generally characteristic of late Ptolemaic and Roman MSS. Ξ is still decidedly Ptolemaic, being formed of three distinct strokes, the middle one of which is very small; so also is Π, which generally retains its breadth. C is well rounded: and the curve of Y is more strongly marked than before. The other letters have no peculiarities that need be noticed. The general aspect of the writing is neat and graceful; less strong than the best third-century hands, but somewhat more ornamental.

[1] *Greek Papyri*, ii. pl. i, no. 2; now Pap. DCLXXXIX *a* in the British Museum.

DIALECTICAL TREATISE.—2ND CENT. B.C.

PLATE XII.

ΜΑΤΩΝ ΑΛΛΕΝΠΡΟCΘΗΚΗCΝΕ ΡΩCΩ
ΔΕΝΟΝΤΑ ΚΑΙCΗΤΩΙ ΑΛΛΩΙΟΦΕΛΕΙΤΗΝΔΑC
ΚΛΙΤΩΝ ΘΡΑΝΩΝΕΙCΜΕΝΟΥΝΔΙΚΑΙΟΚΡΑΤΗC
ΕΝΕΓΕΓΡΑΠΤΟΟΥΗCΕΛΛΟΤΕΛΑ ΡΟΛΥΩΝ
ΟΥΤΟCΜΕΝΕΠΙΤΟΥ ΔΙΟΚΡΑΤΕΥC ΝΟΜΑ
ΤΟ ΗΕΓΡΑΜΜΕΝΟCΟΙΑΛΛΟΙΕΡΗΟΙC
ΕΙΛΗΦΗΠΑΝΤΑΟΜΙΛΛΟΝ ΕΘΟCΥΛΛΟΓΙΙΔΗ
CΙΝ ΟΙ ΤΟΥCΔΟΥΝΕΝΕΓΡΑΤΕΝΕΝΤΑΙC
CΥΝΘΗΚΑΙC ΑΛΛΑΠΕΚΑΤΑΤΟΒΟΥΛΟΥΕ
ΝΟΙCΔΗΜΙΝΕΔΟΧΕΝΠΟΡΕΥΕCΕΛΠΡΟCΕΝ
ΝΚΑΔΙΑΛΕΓΕCΘΑΙ ΑΙ ΛΑΒΟΝ ΘΕ
ΠΤΟΛΠΡΟCΤΟΙCΑΥΓΟΠΩ ΟΙΟΗ ΜΕΝ
ΔΙΟΥΝΑΙCΧΥΝΟΙΤΟΤΕΜΔΟΜΕΝΟCΚΕ ΕΜΥ
ΛΑCΗΜΑCΤΑΙCCΥΝΘΙ ΚΑΙCΟΥΠΡΟCΕΙΙ ΝΤΑ
ΟΔΑΠΕΚΡΙΝΑΤ ΗΜΙΝ ΩCΟ ΓΤΥΤΑ
ΔΙΑΓΗΩCΚΟΙΑΛΕΓΟΜΕΝΟΥ ΤΕΠΙΟC ΝΟΜ
ΤΟΝΝΟΥΝΓΡΑΜΜΑΤΕΙΟΝΤΕΙΗΑ
ΚΕ ΕΝΟΝΠΡΟCΕΜΕΑΡΓΙΤΟΝ ΓΠΟ
ΔΑ ΡΩΠΩΝC ΕΓΟΙ ΝΚΑΡΙΝ
CΗ ΩΤΟΥΠΡ ΙΑΤΟ ΧΙΑ ΗΥΓΟ
ΓΑΙ ΟΥCΛΟΓΟΥCΓΙΝΕ ΕΜΠΩΝ
Τ ΡΤΟΝΚΕΛCΤΟΝ ΟΙ

The Louvre Hyperides[1] presents a very different appearance from its colleague. If the main characteristics of the latter are grace and roundness, those of the Hyperides are strength and squareness. The columns are broad and upright, instead of being narrow and sloping. The letters are firm and noticeably square in appearance, besides being larger than those of the dialectical treatise. The uncial forms of such letters as A, H, M, are retained, and all the characters are fully and carefully formed. This MS. is, in fact, the first example of a phenomenon which runs all through the history of literary papyri, namely the retention of a fully developed uncial form of writing side by side with hands in which a certain concession to the cursive style is discernible. The dialectical treatise is an example of this intermixture of the cursive element, while the Hyperides rejects it. It is obviously more difficult to assign dates to the strict uncial hands, which approach the epigraphic style rather than that of non-literary documents; but there are some letters on which dependence may generally be placed. In the present instance, Ptolemaic characteristics are unmistakably to be found in Λ, the right-hand stroke of which projects very slightly, or not at all, above the point of junction; in M, which has a shallow, angular depression; in the three unconnected strokes of Ξ; and (less distinctively) in the Y-shaped form of Υ. Of the other letters it need only be noted that A and Δ have the peculiarity of a short projecting stroke above the apex; that Z sometimes has its down-stroke drawn to the middle of the lower bar, instead of to its left-hand end; and that ε has the cross-bar rather high, while the upper curve is carried well over, so as nearly or quite

[1] Complete facsimile with printed text, edited by E. Revillout (1893); one column in *Hyperides, the Orations against Athenogenes and Philippides,* edited by F. G. Kenyon; another in Plate XII. Alphabet in Appendix I, col. 5.

to meet it again. The general appearance of the hand is attractive and imposing, and, although less economical of space than the minute hands of the *Phaedo* and *Antiope*, it has good claims to be considered a more handsome style of writing. It is, moreover, very distinct, and therefore a valuable medium for the safe transmission of literary texts.

The Geneva Iliad. Passing from this, the finest example of second-century calligraphy, there are two other literary papyri of the same period which deserve a brief mention. The first of these is a fragment containing a small portion of the eleventh book of the *Iliad*, in which, as in the Petrie fragment mentioned above, there are some additional verses which are not found in the vulgate. The papyrus is at Geneva, and the text has been published by Prof. Nicole, who, however, does not assign any date to it. From a facsimile, however, published by Prof. Diels[1], it is clear that it belongs to the second century B.C., being written, in fact, not in a literary hand at all, but in a reduced form of the common non-literary hand which meets us in the Serapeum papyri[2]. The question as to the character of the enlarged text contained in it does not come within the scope of the present essay; though it may be observed that this is the latest example of such a text which has yet come to light. Some additional specimens of it have lately been acquired by Mr. Grenfell, but they, like the Petrie fragment, belong to the third century B.C.

The Erotic Fragment. The other literary papyrus alluded to above is the fragment of a mime, written either in rhythmical prose or in lyrical verse without strophic correspondences (for scholars differ), recently published by Mr. Grenfell[3]. This

---

[1] *Sitzungsberichte der k. preussischen Akademie*, 1894, no. xix.

[2] It is the more necessary to record the date of this MS., because Prof. Mahaffy, in a moment of aberration, has assigned it to the second century *after* Christ.

[3] *Greek Papyri*, vol. i. (*An Alexandrian Erotic Fragment*), no. 1.   Now Brit. Mus. Pap. DCV *verso*.

unquestionably belongs to the second century B.C., but is written in a non-literary hand, and therefore does not require separate treatment here. It is inscribed on the *verso* of a document dated 173 B.C., in a small cursive hand.

There are other papyri which have sometimes been conjecturally assigned to the second century B.C., notably two of the British Museum MSS. of Hyperides; but in the pages which follow these MSS. have been referred to other periods, and will not be discussed until we arrive at the dates to which, according to the evidence now available, they appear to belong.

From the scantiness of evidence which is thus seen to be available, it is impossible to formularize any very dogmatic conclusions as to the literary palaeography of the second century B.C.; especially as the two MSS. which form the principal sources of our knowledge exhibit very different types of writing. It is fortunate, however, that the approximate dates of these two MSS. are established with more certainty than usual, by the writings which each has upon its *verso*. We are thus in possession of fixed points from which other examples may be dealt with; and in approaching the doubtful period which follows next such fixed points are very valuable.

The first century B.C. is, indeed, a period which requires careful handling and little dogmatism, both because several interesting MSS. have been assigned to it with various degrees of doubtfulness, and because it contains the transition from the Ptolemaic to the Roman periods. A special element of complexity is introduced by the Herculaneum papyri, which, from their character and the circumstances of their preservation, must necessarily belong either to this century or the next, and which, for the first and last time, carry the range of available evidence beyond the borders of Egypt. There are a few fixed points which mark out the stages of the inquiry,

The first century B.C.

and round these the attempt must be made to group the other documents which, on palaeographical grounds alone, appear to belong to the period.

The Her-
culaneum
Papyri.

The Herculaneum papyri, with which it will be convenient to deal first, raise a question which has not yet been discussed, namely how far the history of palaeography in Egypt applies to the Greek-writing world outside. The question cannot be fully answered, for want of sufficient materials; but it can be answered sufficiently for all practical purposes. The analogy of mediaeval Latin MSS. shows that the distinctions between contemporary MSS. written in different countries are less than those between MSS. written in the same country but at the distance of a century or so of time. It requires more experience to determine the country of a MS. than its date. There is no reason to doubt that the same would be the case with Greek writing on papyrus; on the contrary, there is the more reason to expect uniformity, since all Greek MSS., in whatever country, would be written either by Greeks or by those who had learnt their writing from Greeks. We have, in fact, practically the same conditions as prevailed in the Greek minuscule writing of the late Middle Ages. It may be possible for special experience sometimes to distinguish Greek MSS. written in Italy from those written in Greece and Constantinople, but it is not necessary to possess this special knowledge in order to assign the date of a MS. with sufficient accuracy.

When, therefore, we find that the papyri discovered at Herculaneum, though not exactly like any of the Egyptian papyri, yet do not differ from them more than they differ among themselves, and possess strong resemblances both in general appearance and in particular detail, there need be no hesitation in applying to them the same criteria of age as to their Egyptian relatives. No doubt there may have been local distinctions, which fuller evidence might enable us to discern; and certain features may have continued longer or less long in Alexandria or in Rome; but this range of uncertainty is no greater than that which applies to the whole subject of supplying conjectural dates to undated MSS., and in practice may be ignored. It may, however, be worth while to bear in mind that Greek writing was an exotic among the Romans, while it was naturalized in Egypt; hence it is likely that changes in the prevalent styles of writing began rather in Alexandria than in Rome, so that *possibly* MSS. written in Italy may be slightly later in date than MSS. showing the same characteristics written in Egypt. Since, however, the copyists of Greek manuscripts written in Italy were almost certainly Greeks themselves, not much weight need be attached to this argument.

It seems, therefore, perfectly legitimate to use the Herculaneum papyri as evidence for the palaeography of MSS. written in Egypt; and the point is of considerable importance, since the date of the Herculaneum volumes can be fixed with fair accuracy. The *terminus ante quem* is absolute, being the eruption of Vesuvius in A.D. 79, which overwhelmed the town and calcined the papyri. The *terminus a quo* is supplied by the fact that many of the papyri contain works of the Epicurean philosopher, Philodemus, who was a contemporary of Cicero; while nearly all the rest are either copies of the works of Epicurus or treatises of a philosophical character. Now Philodemus was not a very eminent philosopher, and it is not likely that a large collection of his works would be made at any time much after his death; and the fact that several of his treatises are here represented in duplicate suggests the probability that the collection was made by Philodemus himself [1]. This

[1] See Scott, *Fragmenta Herculanensia*, pp. 11, 12.

would assign the MSS. to the first century B.C., and
to dates before, rather than after, B.C. 50. This conclusion is not invalidated by the fact that a Latin
poem on the battle of Actium and the Egyptian campaign
of Augustus was found along with the Greek papyri, since it
may easily have been added by a later owner of the library.
On the other hand, this poem is itself not at all likely to
have been written later than the reign of Augustus, and
therefore supplies a proof that the library was not of
recent formation at the time of the catastrophe of A.D. 79.
A philosophical library of the preceding century may
very well have been kept together in the house of its
original owner, just as libraries of eighteenth-century
theology may be found in country houses nowadays,
even though some of the authors are somewhat out of
date. There is, therefore, no ground for rejecting the
natural inference from the character of the library itself,
to the effect that the volumes composing it were written
shortly before the middle of the first century B. C.

The Herculaneum papyri consequently belong to the
end of the Ptolemaic period of Greek palaeography; for
it will be convenient to retain the name, although it is
something of a misnomer when applied to MSS. written
outside Egypt. Indeed, as has been briefly indicated in
an earlier chapter, the names 'Ptolemaic,' 'Roman,'
'Byzantine,' must be applied much less rigidly to literary
than to non-literary papyri. Changes of dynasty, implying changes of influence and the introduction of clerks
and officials of different origins, may very naturally affect
the handwriting of official, and through them of private,
documents to a very noticeable extent; but the writers
of books belong to a society which extends beyond the
borders of a single kingdom, and would not consciously
submit to the influence of an official chancellery. Hence,
on the one hand, the changes in literary papyri at the
advent of the Romans in Egypt are less obviously recog-

nizable; on the other hand, they may be taken as applying generally to Athens and Rome as well as to Alexandria. In speaking of any Egyptian papyrus, written in a formal book-hand, as 'Ptolemaic,' we mean only that it is written in a hand prevalent in Greek literary circles during the period when the Ptolemies ruled in Egypt, with, perhaps, some slight local modifications due to the fact that the scribe was writing in Egypt; and similarly, in speaking of papyri written at Herculaneum as Ptolemaic, we mean that they are written in the Greek book-hand prevalent during this same period, even though the slight local influences were in this case Italian and not Egyptian.

The conclusion that the Herculaneum papyri are to be referred to the end of the Ptolemaic period suits perfectly with what we know from other sources of the course of palaeographical development, and they thus drop easily into a natural place in the sequence. In all crucial details they answer to the criteria which serve to distinguish Ptolemaic from Roman papyri; and the history of Greek palaeography remains natural and intelligible on this theory, which it is not if they are referred to a date a century or more later. The two letters which are of most decisive importance at this stage are A and Ξ. The uncial form of the former letter (A) does, indeed, as has been stated above, run through all periods, and is found alike in Ptolemaic and Roman MSS.; but the minuscule form is different in the two periods. In literary papyri of the Ptolemaic age the right-hand oblique stroke is always formed separately from the rest of the letter, which is written without lifting the pen, having either an acute angle or a loop in the left-hand corner (Λ, λ, ᴧ, A). In Roman papyri, on the other hand, the A is almost always written in one piece, just as in cursive MSS. (ᴧ, ᴧ). There are exceptions, as in the case of the Bankes Homer (Brit. Mus. Pap. CXIV); but they are few, and the rule will generally be found serviceable, except in the cases where the uncial form of

Test letters : A and Ξ.

the letter is retained. The second test letter is still more valuable, being liable to no such exception as that just mentioned. It is, of course, rash to rely absolutely on the form of a single letter as an infallible test of date, especially where evidence is so scanty as it is in the case of literary papyri; but so far as the evidence goes at present, Ξ in Ptolemaic MSS. is invariably formed of three disconnected strokes [1], while in Roman MSS. it is equally invariably formed in one continuous whole. This rule at present holds good absolutely, though in paleographical matters isolated exceptions to any rule may always come to light.

According to both these tests, the Herculaneum papyri fall within the Ptolemaic period. They include, as might be expected, several different hands, but most of them exhibit the same general type [2]; and in respect of these special letters there is little variation. A is normally of the minuscule type, formed in two strokes; in a few cases, such as the MS. of Philodemus *De Ira*, it is of the uncial shape; but never, apparently, of the Roman type, written without lifting the pen at all. The evidence of Ξ is equally clear and unanimous, the letter being regularly formed of three distinct strokes. The only variation is that in some cases the central stroke is a horizontal line (shorter than the two others), while elsewhere it has the shape of a comma (Ξ, Ⲭ). In one MS.[3], if the facsimile (which is not good) is to be trusted, the central stroke does indeed touch and connect the two others (Ⲭ); but the exception is more apparent than real, since the letter

---

[1] One of Mr. Grenfell's fragments, described above (p. 45) has the three strokes connected by a perpendicular line at right angles to them (Ⲭ); but this is a still more archaic form, and does not in any way invalidate the principle here enunciated.

[2] See facsimiles in Scott's *Fragmenta Herculanensia*; *Thirty-six Engravings of Texts and Alphabets from the Herculaneum Fragments* (ed. Nicholson, Oxford 1891); *Pal. Soc.* i. 151, 152; and a series of photographs issued by the Oxford Philological Society, in several volumes. Alphabet in App. I, col. 7, below.

[3] Scott's *Pap.* 26.

is still formed by three distinct strokes of the pen, the
only difference being that the central comma-shaped stroke
is large enough to touch both the upper and the lower
horizontal line.

In general appearance the Herculaneum hands are
small and graceful, having affinities with the Louvre
dialectical fragment described above, rather than with the
larger and squarer type of the Louvre Hyperides. A, as
will be seen from the above description, is not unlike the
same letter in the dialectical papyrus; Є is generally well
rounded, and the central stroke has a tendency to be
separated from the rest of the letter; the two strokes
of Λ are generally united at the top, but sometimes the
right-hand stroke is prolonged; M is Ptolemaic, having
an angular centre instead of the deep rounded curves of
the Roman period; Π is fairly broad; Υ is of the Υ-shape,
but without any strong individuality. All these details
tend to confirm the assignment of these papyri to the
end of the Ptolemaic period, and assist in giving fullness
and precision to our knowledge of its palaeographical
characteristics.

The Herculaneum papyri having thus been assigned
to their place, two other MSS. must be mentioned which
likewise seem to belong to the earlier half of the first
century B. C. The first of these is the recently discovered
MS. containing the poems of Bacchylides[1]. This has
strong Ptolemaic features; but something in its general
appearance, as well as some details which it has in
common with later MSS., indicates that it probably
belongs to the end of the period, about the middle of the
first century B. C. It is in a handsome hand, of good size,
clear and firm, and carefully written. It evidently belongs
to the class of MSS. written for sale or for preservation

The Bacchylides Papyrus.

---

[1] Complete facsimile published by the Trustees of the British Museum.
One column is reproduced in Plate XIII. and an alphabet given in App. I,
col. 6.

in a public library, and ranks high among the extant
specimens of Greek calligraphy upon papyrus. Its columns
are wide (about 5 inches), their size being determined by
the length of the longest of the irregular verses in which
the odes are written. Of the individual letters, A is conspicuously angular, without a vestige of a curve, and
formed in two strokes of the pen, (A, Λ); ε is narrow,
with a long projecting central stroke; Z has the peculiarity
of often having the lower horizontal stroke separate from
the rest of the letter; Λ shows a slight projection of the
right-hand stroke, which may be taken as an indication
of the approach of the Roman period; M, on the other hand,
is decidedly Ptolemaic, being a broad letter with a shallow
curve; Ξ, too, is quite Ptolemaic, and the central bar is
generally reduced to a mere point, while the other strokes
are unusually long; O is small; Π is broad; C is narrow,
and the upper part of it is often separated from the rest,
a phenomenon which recurs in the Harris Homer, to be
described in the next chapter; T has its cross-stroke longer
on the left than on the right; Y has a very shallow curve;
ω is broad and shallow, with little or no traces of the
central stroke. Some of the titles of poems have been
added in a different hand, which is plainly of Roman
type, and probably not earlier than the second century;
but these are so clearly a later addition, being omitted in
some instances, and showing quite distinct forms of letters,
that they give no clue to the date of the original writing [1].

---

[1] The date here assigned to the Bacchylides papyrus is questioned
by Messrs. Grenfell and Hunt *Oxyrhynchus Papyri*, i. 53), who would place
it in the first or second century after Christ, partly on the strength
of a papyrus of Demosthenes published by them, which (having accounts
on the *verso* in a second-century hand   they assign to the early part of the
second century.   The resemblance between that papyrus and the Bacchylides, however, is not very great, and to my eye the Demosthenes appears
decidedly later.   None of the most characteristic letters of the Bacchylides
(μ, ξ, υ, ω has the same shape in the Demosthenes.   Personally I should
put the Demosthenes fragment somewhat earlier than they do, regarding
the cursive writing on the *verso* as of the middle of the second century,

PLATE XIII.

ϹΠΙΒΑΝΙΑΠΟΛΑΜΝΙΩΝ
ΦΟΙΝΙϹϹΩΦΙΛΟ ΓΑΛΑΧΩΝΔΕΥΝ
ΠΡΩΘΗΔΟΝΑ ΡΑΙΩΝΟΘΛΥΕΛΑΤΕΝ

ΚΗΤΤΥΚΤΟΝΚΙΝΕΟΝΔΑΚΝ
ΝΔΝΚΡΔΤΟΙΓΠΕΡΠΤΡΟΟΧΑΙΤΑΤ
ΧΙΤΩΝΑΠΟΡΦΥΡΕΩΝ
ϹΤΕΡΝΟΙϹΠΑΜΦΙΚΛΙΝΟΤΝΟΗ
ΘΕϹϹΑΛΑΝΧΛΑΜΥΔΟΜΜΑΤΩΝΙΟΙ
ΜΕΗΝΔϹΘΑΙΠΟΛΕΜΟΥΤΤΕΚΩ
ΧΑΛΚΕΗΚΤΥΠΟΤΗΑΧΩΝ
ΔΙΖΗϹΘΑΙΟΑϹΦΙΔΛΑΛΟΥϹ ΜΑΝ ΝΑ
ΠΑΡΕϹΤΙΜΥΓΓΙΑΚΕΛΕΥΘΟΙ
ΑΜΒΡΟϹΙΩΝΜΕΛΕΩΝ
ΟϹΑΝΠΑΡΑΠΙΕΡΙΔΩΝΛΑ
ΧΗΙϹΙΔΩΡΑΜΟΥϹΑΝ
ΙΟΒΛΕΦΑΡΟΙΤΕΚΩ
ΦΕΡΕϹΤΕΦΑΝΟΙΧΑΡΙΤΕϹ
ΒΑΛΩϹΙΝΑΜΦΙΤΙΜΑΝ
ΥΜΝΟΙϹΙΝΥΦΑΙΝΕΝΤΝΝΕΝ
ΤΑΝΕΠΙϹΧΗΡΙΑΤΟΙϹΤΙΚΑΝΟΝ
ΟΛΒΙΔΝϹΑΘΑΝΑΙϹ
ΕΤΑΝΗΕΤΕΚΗΙΔΜΕΡΙΜΝΑ
ΠΡΕΠΕΙϹΕΦΕΡΤΑΤΑΝΙΜΕΝ
ΟΔΟΝΠΑΡΑΚΑΛΛΙΟΠΑϹΑΝ
ΧΟΙϹΑΝ ΕΞΟΧΟΝΓΕΡΑϹ
ΤΙΗΝΔΡΓΟϹΟΦΙΠΠΙΟΝΛΙΠΟΥϹΑ
ΦΕΥΓΕΧΡΥϹΕΑΒΟΥϹ
ΕΥΡΥϹΘΕΝΕΟϹΦΡΑΔΑΙϹΙΦΕΡΓΟϹΤΑΛΟΙ
ΙΝΑΧΟΥΡΟΔΟΔΑΝϹΤΡΟΙ ΚΩΡΑ
ΟΤΑΥΓΟΝΟΜΗΑϹΙΒΛΕΠΟΝΤΑ
ΠΑΝΤΟΘΕΝΑΚΑΜΑΤΟΙϹ
ΜΕΓΙϹΤΟΘΕΝΔΙϹΙΔΝΚΔΔΥΓΕΝ

The remaining MS. to be noticed is that which contains the oration of Hyperides against Philippides and the third Epistle of Demosthenes (Brit. Mus. Papp. CXXXIII, CXXXIV)[1]. These two compositions, written on a single roll of papyrus, are in quite different hands, both of them rather peculiar and noticeable. The Hyperides is written in small, very graceful letters, well-rounded, with ligatures as distinctly marked and carefully written as the letters themselves, and in exceptionally narrow columns. The Demosthenes is in yet smaller letters—indeed there is no other literary

Hyperides In Philippidem.

and the Demosthenes as of the first century; which would allow of the Bacchylides remaining where I have placed it above, at the transition from the Ptolemaic to the Roman style in the first century B. C. Messrs. Grenfell and Hunt further state that the forms of μ and ν which appear in the Bacchylides are also found in two of their papyri of the Roman period, the Aristoxenus and Thucydides fragments; but the μ in both these MSS. is quite different, being of only ordinary breadth and much more deeply indented, and the ν, though shallow in the top, does not very closely resemble the same letter in the Bacchylides. On the whole, the Oxyrhynchus papyri, which are all of the Roman period, seem to me to confirm the date here assigned to the Bacchylides. Prof. Blass (*Bacchylidis Carmina*, 1898, pp. vii, viii) assigns the Bacchylides MS. to the first century after Christ rather than to the previous century, on grounds not of palaeography but of orthography, especially its comparative freedom from iotacism. His argument is that (according to the evidence of inscriptions) under the earliest emperors the interchange of $\epsilon\iota$ and ι was carried to an extreme, while in the course of the first century a reform in this respect was introduced, culminating in the second century under the influence of Herodian. In the absence, however, of more manuscript evidence on the point than is at present available, this argument is somewhat precarious. It is certain that in non-literary papyri the interchange of $\epsilon\iota$ and ι is quite as common in the first and second centuries after Christ as in the Ptolemaic period, and it is clear that much must have depended on the practice of individual scribes. A well-written MS., such as the Bacchylides, copied from an earlier MS. free from iotacism, would not be likely to be affected by it to any serious extent, certainly not to any extent comparable to inscriptions or non-literary papyri in which the scribe had no earlier copy to follow. Under these circumstances, the evidence of orthography seems much too doubtful to be accepted as the main guide to the date of a MS. : and the fact that this evidence leads Prof. Blass to assign the Herodas papyrus to the Ptolemaic age may be taken as reinforcing this conclusion (see below, pp. 94, 95).

[1] Specimen facsimiles in *Classical Texts from Papyri in the British Museum*, ed. F. G. Kenyon (1891), plates ii and iii. Alphabets in App. I, cols. 7 and 8.

papyrus in which the characters are so small—more angular and less well formed. The columns of both lean decidedly to the right. In detail the letters are akin to those of the Herculaneum papyri, and it is on the strength of this comparison that the MS. is assigned to the first century B.C. In the Hyperides, A has the loop in the left-hand corner which has been noticed in the Herculaneum papyri (and in an incipient form in the Louvre dialectical fragment), the right-hand stroke being made separately; Δ is formed in a very similar way, but the loop does not fall below the rest of the letter; Є is well rounded, with the cross-stroke rather high; Λ shows some projection of the right-hand stroke; K is deeply curved, a sign of the approach of the Roman style; Ξ is Ptolemaic, with three distinct strokes; C is well rounded; Y is deeply curved, much as in the reign of Augustus; ω is fully and carefully formed. The Demosthenes has many of its details very similar, but is less well written throughout. A varies between the uncial form and the form with the loop in the left-hand corner; some shapes of the latter show the approach of the Roman type, but the letter is still written in two strokes; Є is small and narrow; Λ sometimes has the right-hand stroke projecting, sometimes not; M is generally shallow, recalling the Ptolemaic type; Ξ has its three separate strokes; C has rather a flat top; Y resembles the same letter in the Bacchylides papyrus; and ω is only slightly formed, the second loop being often hardly visible. It will be seen from a consideration of these features, and especially of such test letters as A, M, and Ξ, that both hands of this MS. are on the border line between the Ptolemaic and the Roman periods. Some forms are still distinctly Ptolemaic, while others are already approaching the types which will be seen in the next chapter to belong to the Roman style. They stand or fall with the papyri of Herculaneum, and cannot be placed very far from them in point of time.

Demo-
sthenes,
Ep. III.

With these MSS., therefore, we reach the end of the Ptolemaic period, which may be roughly fixed at the middle of the first century B.C. Such precise definitions of time are, however, only useful as aids to the memory, and it is as well to repeat the caution that no precise accuracy can be expected in the dating of literary papyri in the present state of our knowledge. Palaeographical periods necessarily melt into one another, and the forms of letters have no precise boundaries. Even where materials are much more plentiful, as in the minuscule MSS. of the twelfth to the fifteenth centuries, it is only necessary to look at the published series of dated facsimiles from MSS. in the Bibliothèque Nationale, to have any rash dogmatism checked by the recognition of the fact that types of hands run on much further than is commonly allowed for. Dating MSS. by palaeographical indications alone is, to a considerable extent, a science of conventions. We can say that such and such a MS. is written in a hand which was prevalent in, say, the first century; but we can much less confidently affirm that it was itself certainly written in that century. Still, it is only by provisional generalizations that science progresses; and this is as true of palaeography as of physics or chemistry. The crystallization of isolated observations into formulas may lead to some mistakes, but it also leads to progress, by the confirmation or rejection of the formulas; and so this mapping out of the province of Ptolemaic palaeography will, it may fairly be hoped, be of some use for the arrangement and systematization of existing knowledge, and as an aid towards the assimilation of the knowledge which future discoveries may bring in. The criteria which have been indicated in the course of this chapter meet, it is believed, all the data which are now forthcoming, and arrange them in an intelligible series; but their final verification, or otherwise, must come from the evidence which the spade of the explorer has still to bring to light.

CHAPTER V

THE Roman period is, in relation to non-literary papyri, the period concerning which we have the fullest and the most precisely dated evidence; but the same cannot be said of it in respect of literary MSS. In mere number, indeed, the literary documents of the Roman period exceed those of the Ptolemaic: but it happens that hardly any of them can be even approximately dated upon independent grounds. Consequently there has been great uncertainty as to the dates assigned to many of the MSS. with which we have to deal in this chapter: and it would be rash to say that the period of uncertainty is yet at an end. Any day some fresh piece of evidence may be brought to light which will modify the conclusions to which we can now come; and all statements which rest solely on the impression produced by a MS. itself, without collateral evidence connecting it with better established knowledge, must be taken as only approximate.

We have seen, at the close of the preceding chapter, the progress of the transition which leads from the Ptolemaic to the Roman age, and have found it exemplified chiefly by the papyri from Herculaneum. Speaking generally, the characteristics which distinguish the Roman period from its predecessor are (1) a greater roundness and smoothness in the forms of letters, and (2) a somewhat larger average size. The very small hands which occur

in so many—and those often the best written—of the
Ptolemaic MSS. do not recur; it is rather the characters of
such papyri as the Louvre Hyperides and the Bacchylides
that set the pattern for the Roman period. And letters
which are prevailingly straight or angular in the earlier
period, such as A, M, Ξ, are now rounded off into curves;
while letters like Є and C, which were rounded before,
become more rounded now.

Before coming to a definitely dated MS. which shows
the characteristics of the Roman style fully established,
there are one or two papyri to be mentioned. in which
the remains of Ptolemaic influence seem to be visible.
One of these is a fragmentary roll of the last two books
of the *Iliad*, now in the British Museum [1]. It is written
in a hand of fair size, somewhat square in shape, with
thin, firm strokes. The ι adscript is regularly written,
which is a sign of a relatively early date: and there are
no breathings, accents, or marks of elision by the first
hand. There are, however, some of the critical symbols
(the διπλῆ and asterisk) employed by Aristarchus, of which
this MS. provides the earliest examples as yet extant.
In form several of the letters show distinctly Roman
characteristics. A is formed in a single stroke, with
a curved body, though there are also signs of the older
angular shape: M is deeply curved; Є and C are well
rounded, the former having the cross-bar high; the right-
hand stroke of Λ projects considerably; and Ξ is of the
Roman type, formed continuously. On the other hand
Y has the shallow top which appears also in the Bacchylides
and Demosthenes; and this, along with the somewhat transitional form of A and the generally early aspect of the
writing as a whole, seems to justify the attribution of the
MS. to the latter half of the first century B. C.

Homer.
Il. xxiii,
xxiv.

[1] Pap. cxxviii; collation, with specimen facsimile, in *Classical Texts from
Greek Papyri*; complete text in *Journal of Philology*, xxi. 296. Alphabet in
App. I, col. 10.

G

The
Louvre
Alcman.

Another MS., of less palaeographical, but greater literary, interest, which has sometimes been assigned to the Ptolemaic period, is the fragment of Alcman, now in the Louvre [1]. As containing some otherwise unknown lines of this lyric poet, it is a papyrus of considerable value: but it is not a regular literary MS., being roughly written in a non-literary hand. This hand is, no doubt, a fairly early one, but there seems to be no reason for placing it earlier than the Christian era. All the test letters are of the rounded Roman type; while Y has the deep curve and the tail bending far to the right, which are characteristic of the non-literary hands of the first half (and especially the first quarter) of the first century. The fragment may be of the reign of Augustus, but, if so, it belongs to the last years of that emperor.

Brit. Mus.
Pap.
CCCLIV.

We are fortunate in possessing a MS. which enables us to fix the exact state of palaeographical development which had been reached at the beginning of the Christian era. This is Pap. CCCLIV of the British Museum, a document of non-literary character, but written in a careful and most elaborate book-hand, and capable of being precisely dated [2]. It is a petition for redress of injuries, addressed to the prefect of Egypt, Gaius Turranius. This officer is known from an inscription (*C. I. G.* iii. 4923), which contains a date in the reign of Augustus; and though the date is somewhat mutilated, and has been differently read by different scholars, it is certain that it corresponds to either 15, 10, or 7 B.C. The difference between these dates is quite immaterial for palaeographical purposes, and we thus have a definitely fixed date for the type of hand shown in this MS. It is a hand of great beauty, of medium size, regularly and firmly written, with

---

[1] Facs. in atlas to *Notices et Extraits*, pl. l.

[2] Facs. in the atlas to vol. ii. of *Greek Papyri in the British Museum*, and a partial facs. (showing best-preserved part) in Plate XIV. Alphabet in App. I, col. 11.

PLATE XIV.

PETITION.— *circ.* B.C. 10.

graceful and well-formed strokes and curves. There are, perhaps, no better written papyri in existence than this and the MS. to be mentioned next, which is closely akin to it. Of the individual letters, A is generally rounded, but with some examples of the angular shape; M is pointed in the middle, which recalls one of the Ptolemaic forms; Ξ is written continuously, but with more sharpness and precision than is usual in later Roman MSS.; Υ is moderately curved, and the tail is usually kept to the right, though not bent away as in the example last quoted; Є, Ο, С, Ѡ, are all well rounded and fully formed: the right-hand stroke of Λ projects upwards.

It so happens that the discovery of this document enables us to assign a date, with approximate accuracy, to a literary papyrus of considerable interest, containing the latter part of the third book of the *Odyssey*[1]. The interest of this MS. lies not only in the fact of its being the earliest extant copy of the *Odyssey* (which is of very much less frequent occurrence in papyri than the *Iliad*), but also in the beauty of the writing, which entitles it to be regarded as the handsomest literary papyrus at present known to exist. A *terminus ante quem* is given by the occurrence of some scholia, written in a cursive hand which appears to belong to the end of the first century; but a more precise date may be derived from its close resemblance in general appearance to the MS. which has just been described. The resemblance, which cannot be missed by any one who compares the two documents, renders it certain, even apart from the collateral evidence furnished by the scholia, that the Homer was written at no great distance of time from the petition to Turranius; while an examination of details shows that the latter is the earlier of the two. In the Homer the A is more rounded, M is deeply curved, Ξ is written continuously,

The British Museum *Odyssey*.

---

[1] Brit. Mus. Pap. CCLXXI. Facs. in *Pal. Soc.* ii. 182, partial facs. in Plate XV; text in *Journal of Philology*, xxii. 238. Alphabet in App. I, col. 12.

and curves replace the sharp angles noticed in the petition. In all these typical letters the angular shapes have given way to curves which are characteristically Roman in nature. Of the other letters, Є and С are well rounded, the former having the cross-stroke very high, much as in the *Iliad* MS. last described; H has the cross-bar exceptionally high; and Y approaches the y-shape, the two upper limbs branching off more equally than has usually been the case hitherto. This is a transition shape in the direction of the V-form which prevails later. The general conclusion, therefore, to which palaeographical data point is that the *Odyssey* papyrus was written near the beginning of the first century, and that the scholia, which are of later appearance and include quotations from the grammarian Apion, who flourished under Caligula, were added in the second half of the same century.

The Harris Homer.

More doubt attaches to another Homer papyrus, containing the eighteenth book of the *Iliad*, and commonly known (from the name of its first purchaser) as the Harris Homer[1]. This was one of the first literary papyri to be brought to light, and consequently enjoyed a somewhat exaggerated reputation of extreme old age. Sir E. Thompson, whose authority in such matters is of the highest, places it 'without hesitation as early as the first century B. C.[2]' The increase of evidence during recent years makes it very doubtful whether so early a date can be maintained. The whole aspect of the hand is Roman, not Ptolemaic, and one might even be inclined at first sight to refer it to the second century, were it not for certain letters which still recall the hands of the first century B. C. These are especially A, which is of

[1] Brit. Mus. Pap. cvii; facs. in *Pal. Soc.* ii. 64, and a less successful one in the *Cat. of Anc. MSS.* (Greek). Alphabet in App. I, col. 13.

[2] *Handbook*, p. 124. In the *Pal. Soc.* and *Cat. of Anc. MSS.* it is simply described as 'perhaps first century B. C.'; but the *Handbook* is the latest statement.

PLATE XV.

ΛΛΟΝΓΑΡΑΤΑΛΛΑΚΝΙΜΦΟΒΕΛΟΙΟΙΝ
ΟΝΖΛΚΡΟΠΟΡΟΥΟΟΒΕΛΟΥΟΕΝΧΕΡ
ΛΕΤΗΛΕΜΑΧΟΝΛΟΥΟΕΝΚΛΛΗΠΟΛΥ
ΕΓΟΛΛΟΤΑΤΗΘΥΓΑΤΗΡΗΛΗ
ΕΠΕΙΛΟΥΟΕΝΤΕΚΑΙΕΧΙΤΕΝΛΙΠΕ
ΛΕΝΙΝΦΛΡΟΟΚΛΛΟΝΒΛΛΕΝΗΔΕΧΗΤ
ΤΙΟΛΜΙΝΘΟΥΒΗΔΕΜΛΟΛΘΛΝΑΤΟΙΟΙΝ
ΔΟΓΕΝΟΟΤΟΡΙΩΝΚΑΤΑΡΕΧΕΤΟΤΙΟΙΜ
ΟΙΛΟΤΕΙΩΠΗΤΙΟΙΝΚΡΕΥΠΕΡΤΕΡΑΚΛΙ
ΔΗΝΥΝΘΕΧΟΜΕΝΟΙΕΠΙΛΛΝΕΡΕΟΕΟΘΛ
ΟΙΝΟΝΕΝΟΙΝΟΧΟΕΥΝΤΕΟΕΝΙΧΡΥΟΕΟΙΟ
ΛΥΤΛΙΕΠΕΙΠΟΟΙΟΟΚΛΙΕΔΗΤΥΟΟΕΞΕΡΟΝ
ΤΟΙΟΙΔΕΜΥΘΩΝΗΡΧΕΟΕΡΗΝΙΟΟΙΠΠΟΤΛ
ΤΗΛΔΕΟΕΜΟΙΛΓΕΤΗΛΕΜΛΧΩΙΚΛΛΗΠΤΙΧΛΟΙ
ΖΕΥΞΛΘΥΦΛΡΜΑΤΑΓΟΝΤΕΟΙΝΛΠΗΤΟΟΜΟΙΝ
ΩΟΕΦΛΘΟΛΡΑΤΟΥΜΑΛΛΕΝΚΛΥΟΝΗΔΕΓ
ΚΑΡΠΑΛΙΜΩΟΔΕΖΕΥΞΛΝΥΦΛΡΜΑΟΙΝΩΚΕΛ
ΑΝΔΕΓΥΝΗΤΑΜΙΗΟΙΤΟΝΚΛΙΟΙΝΟΝΕΘΗΚ
ΟΨΛΤΕΟΛΡΛΟΥΟΙΛΝΟΤΡΕΦΕΕΟΒΛΟΙΛΗΕΟ
ΑΝ ΔΛΛΡΤΙΛΕΜΛΧΟΟΠΕΡΙΚΛΛΛΕΒΗΟΕΤΟΛ
ΠΛΡΔΛΙΝΕΟΤΟΡΙΛΠΟΠΙΟΤΟΙΟΤΡΑΤΟΟΟΡΧΛ
ΕΟΧΙΦΡΟΝΛΧΝΕΒΛΙΝΕΚΛΙΗΝΙΛΛΛΧΕΤΟΧΕ
ΜΛΟΤΙΞΕΝΛΕΚΛΛΝΤΩΛΘΥΚΤΛ.ΩΝΤΕΠΟΗ
ΕΟΤΕΛΛΟΜΛΙΠΕΤΗΝΔΕΠΥΛΟΥΛΙΠΥΠΤΟΛ
ΤΛΝΗΜΕΡΙΟΙΟΕΙΩΝΧΥΓΟΝΛΜΦΙΟΛΕΧΟ
ΒΗΡΑΟΛΙΚΟΝΤΟΔΙΟΚΛΗΟΟΓΟΗΛΟΟΛΙΗ
ΟΟΡΤΙΛΟΧΟΙΟΤΟΜΑΛΦΕΙΟΟΤΕΚΕΤΙΛΙΝΛ
ΕΝΥΚΤΛΕΟΛΝΟΛΕΤΟΡΛΠΛΖΕΙΝΙΛΘΗ
ΗΜΟΟΛΗΡΙΓΕΝΕΙΛΦΛΝΗΡΟΔΛΟΛΛΚΤΥΛΟΟ
ΙΠΠΟΥΟΤΕΧΕΥΓΝΥΝΤΛΝΛΘΟΛΕΜΛΤΑΠΟΙΚΙΛΛ

the angular type which is found in the Bacchylides, though
somewhat less sharply pronounced; and Є, which, again
as in the Bacchylides, is narrow, with a long projecting
cross-bar, and with the head of the letter often separated
from the rest. On the other hand M and Ξ are distinctly
Roman in character; Λ always has a projecting right-hand
stroke; M, N, Π, are square, not elongated as in Ptolemaic
hands; С has a somewhat flattened top; Y is approaching
the V pattern, and has quite lost its Ptolemaic appearance.
Some omitted lines have been supplied in the margin in
an ordinary cursive hand which seems to belong to the
beginning of the second century; and it is probably the
same hand that has added the accents and breathings.
On the whole it would appear that the MS. must be
assigned to the first century of our era; though of course,
here as always, no pretence of dogmatism is admissible,
especially in conflict with other authorities.

Another MS. which has commonly been placed in the
first century B. C., and sometimes even earlier, but which
may now almost certainly be brought down to the first
century after Christ, is the great papyrus of Hyperides—
the one by which his work was first made known to the
modern world—containing the prosecution of Demosthenes
and the defences of Lycophron and Euxenippus—the
latter being the only extant speech of the orator which
is absolutely intact. At the date when this MS. was
acquired by Messrs. Harris and Arden, there was practically
no means of forming an accurate idea of its age, and its
first editor, Mr. Babington, cautiously placed it between
the second century before, and the second century after,
the beginning of the Christian era. Subsequent scholars,
who have tried to be more precise, have in fact ranged
over the whole of this somewhat wide period. Sir
E. Thompson has assigned it to the first century B. C.[1];

The British Museum Hyperides.

[1] *Cat. of Anc. MSS. (Greek)*; *Handbook*, p. 123. A complete facsimile
of the MS. (by hand) is given in Babington's edition; photographic

Prof. Blass to the second century after Christ[1]. But this discrepancy between two eminent palaeographers is due to the fact that they based their conclusions upon different evidence. Prof. Blass was guided by the cursive hand in which the title of the MS., giving a list of the orations contained in it, is written, which he rightly assigned to the second century after Christ; while Sir E. Thompson equally rightly argued that the title is not in the same hand as the body of the MS., and therefore gives nothing more than a *terminus ante quem* for our assistance. It is, however, now sufficiently clear that no date within the Ptolemaic age is possible for this style of writing. The most marked feature of the hand in its general appearance is its roundness—a distinctively Roman characteristic; and the forms of the single letters all point the same way. A is written continuously, with rounded angles and somewhat loose construction; M is fully rounded; Ξ is continuous, and rather irregular; C is rounded, and even shows a decided tendency to fall forward—an indication of a date nearer the end of the first century than the beginning; Y is of the characteristically Roman type. The letters are not always well or firmly formed; and that again is a sign of a relatively late date.

On these grounds alone it would seem almost certain that this papyrus cannot be placed earlier than the second half of the first century; and this conclusion has been strongly confirmed by evidence which has lately come to light. In the first place, striking resemblances to this Hyperides MS. are found in the long poll-tax rolls recently acquired by the British Museum, which bear dates in the years 72 and 94[2]. Secondly, a small scrap

facsimiles of parts of it in *Pal. Soc.* i. 126, and *Cat. of Anc. MSS.* Some recently discovered fragments, now in the possession of Rossall School, are shown in Plate XVI, through the kind permission of the Rev. J. P. Way, head-master of Rossall. Alphabet in App. I, col. 14.

[1] I. von Müller's *Handbuch der klassischen Altertums-Wissenschaft*, ed. 2, p. 312.

[2] Brit. Mus. Papp. CCLVII-CCLIX, CCLX-CCLXI.

HYPERIDES IN DEMOSTHENEM AND DEMOSTHENES OLYNTHIAC II.—LATE 1ST CENT.

of a literary papyrus, in a hand almost identical with
the Hyperides, but somewhat larger, was acquired along
with a number of dated documents of the first and second
centuries[1]. But the most conclusive evidence is afforded
by a fragment brought from Egypt by Mr. Grenfell
in 1895, and now in the Bodleian Library. It contains
a small portion of the *Homeric Lexicon* of Apollonius,
though in a form somewhat different from that in which
it has hitherto been known[2]. The evidence for its date
is a little complicated, but seems fairly conclusive, at
least as regards the *terminus a quo*. The grammarian
Apion is mentioned in various places in the *Lexicon*, and
one of these passages occurs in the Bodleian fragment.
The papyrus is, however, mutilated in this place, and
the name of Apion is not preserved; so that it is con-
ceivable—considering the extent to which the texts of
Apollonius hitherto extant have been expanded from the
original—that the name might be a later addition. But
the size of the lacuna just suits the supposition that the
papyrus here had the same text as the later copies, and
there is no other natural way of filling it. It is therefore
a legitimate conclusion that Apion was mentioned in the
papyrus, when it was perfect; and since Apion flourished
under Tiberius and Caligula, it is practically certain that
the papyrus cannot be earlier than the middle of the
first century, while it may, of course, so far as this argument goes, be considerably later. The importance of this
conclusion lies in the fact that the fragment is written
in a hand of precisely the same type as the long
Hyperides MS.; and if the Apollonius is not earlier
than the middle of the first century, the same must also
be the case with the Hyperides. The likeness between
the two MSS. is as great as it can be between two

The Bod-
leian
Apollo-
nius.

[1] Brit. Mus. Pap. ccviii c.
[2] A facsimile has been privately circulated by Bodley's Librarian, to
whom the identification of the fragment is due.

hands which are not identical, and will be disputed by no one who has compared the two.

It seems established, therefore, that the theories which assigned the Hyperides to the Ptolemaic age, or even to the first century of the Roman period, must be abandoned : but there still remains the possibility that Blass is right in placing it as late as the second century. But, in the first place, the main ground of his view—namely, that the title of the MS. is written in a cursive hand of the second century—is unsound ; for the date of the title does not necessarily establish the date of the MS., but merely gives a *terminus ante quem*. Titles not only might be, but actually were, written much later than the MSS. to which they applied. A striking instance is given by the Bacchylides papyrus, in which some of the poems are without titles, while those which exist have plainly been written at a later date than the poems themselves. So far, then, as the evidence of the title goes, all that can be said is that the MS. cannot be later than the end of the second century, though it may be considerably earlier. It is here that the evidence of the poll-tax rolls, above mentioned, becomes important ; for they show that hands of the same class as that of the Hyperides were in use in the last thirty years of the first century, while no example of such hands has yet been found of a later date. To this period, then, of the end of the first century, the great Hyperides MS. must almost certainly be referred. External testimony here strongly supports the indications of palaeographical science, and it will need clear evidence to overthrow their joint conclusion.

Brit. Mus. Pap. cxli.   At this point in the history of papyrus-palaeography we come upon another document which, though not literary in its contents, is yet, like the petition to Turranius of *circ.* B.C. 10, of considerable value for the study of literary hands. It is a lease of an oliveyard and other lands in the village of Soenopaei Nesus, dated in the year

PLATE XVII.

LEASE.—A.D. 88.

88; and instead of being written, as usual, in a small and very cursive hand, it is executed in careful uncials of a large size, with only a sprinkling of minuscule forms [1]. In size, the characters of this MS. are larger than any that we have yet described, though some later documents perhaps equal them. In type, they are plainly the forerunners of the handsome uncials in which the earlier vellum MSS. are written. Before the discovery of this papyrus, with its precise date near the end of the first century, there was little or nothing to show that such hands went back earlier than the fourth century at most. The present document, though by no means perfect in execution, shows that we must look much higher for the origin of this hand; and it also suggests a conclusion to which we shall have to refer again presently, that the palaeography of Greek papyri anticipated in its development the subsequent history of writing upon vellum, so that the corresponding stages of writing on the two materials are not contemporary, but are separated by some centuries of time.

The letters of this papyrus are not uniformly uncial in character, some of them being merely the current minuscule forms written on a large scale. This is notably the case with A and Y, while Є, which is correctly formed as a general rule, occasionally lapses, through the forgetfulness of the scribe, into a completely minuscule form. Of the other letters, M is conspicuously shallow in its central depression; Ξ is of the regular Roman type, and somewhat cursive in appearance; C is well formed and upright; O is large and prominent; Ѡ is well written. The MS. as a whole must be regarded, not as a thorough-going example of a book-hand of a large uncial type, but as an indication of what the contemporary book-hand might be. We cannot say that this is a hand of the same character as those

---

[1] Facsimile in *Pal. Soc.* ii. 146, and in the atlas to *Greek Papyri in the British Museum*, vol. ii; partial facs. in Plate XVII. The papyrus is known as Brit. Mus. Pap. cxli.

which are found in the great vellum MSS. of the fourth and fifth centuries; but we can say that, if this hand could be found in a non-literary document of the year 88, then a hand of the same character as those of the vellum MSS. may have been in use for literary papyri at the end of the first century. The large uncial hand, in short, instead of being a creature of the fourth century, called forth by the adoption of vellum into general use, exists on papyrus at least three centuries earlier.

Brit. Mus. Pap. CLXXXVII. Further evidence in the same direction is given by another papyrus in the British Museum (Pap. CLXXXVII), which contains a small fragment of a moral or historical treatise, giving a description of the training of the youths of some nation, apparently the Lacedaemonians[1]. Only some sixteen lines are preserved intact, or approximately so, with small traces of some other lines in the same and adjoining columns; but quite enough remains to show that it was a carefully written MS., in a large uncial hand. The letters are of about the same size as those of the lease just described, and are free from the intermixture of minuscules by which that is disfigured. Its date is not known with certainty, but the evidence which has just been given authorizes us to place it at the end of the first or in the course of the second century; and this is confirmed by the fact that the *verso* of the papyrus bears writing in a cursive hand which may safely be referred to the beginning of the third century. The writing of the Λακεδαιμονίων πολιτεία (or whatever the treatise is) is careful, delicate, and handsome; and its size and shape put it definitely in the line of ancestry of the vellum uncials. A continues to be of the rounded form; M is very deeply curved; C is so much rounded as to be hardly distinguishable from O, a feature which is perhaps in favour of a date in the second rather than the first century. The

---

[1] Published (without facsimile) in the *Revue de Philologie*, **xxi.** 1 (1897).

columns of writing are narrow, containing, in this large hand, only some twelve to fifteen letters.

Between the first and the second centuries after Christ it is impossible, at any rate with the evidence now available, to draw any firm line of palaeographical demarcation. No external change came over the government of Egypt which could affect the local handwriting. The development of the Roman hand pursued a natural course; and in the absence of precisely dated examples it is impossible to fix its stages with certainty. There are a considerable number of MSS. which appear to belong to this period; but it would require a bold palaeographer to decide, with regard to most of them, whether they were written in the second century or the last half of the first. It so happens that most of the MSS. referred to contain parts of Homer; but the list also includes the unique papyrus of Herodas, a speech of Isocrates, and a few minor pieces[1].

Close to the border line, but almost certainly within the limits of the first century, comes a papyrus of very great literary interest, namely that which contains the Ἀθηναίων Πολιτεία of Aristotle[2]; but unfortunately, of the four hands in which it is written, only one at all approaches the character of a book-hand, and that is plainly not the work of a trained scribe, from the number of clerical blunders which it contains. Rather is the MS. a striking example of what may be called the private method of circulating literature in the ancient world. Libraries were few and far between, and copies of literary works written by professional scribes may have been difficult and expensive to obtain in the upper parts of Egypt. Hence we seem to see the existence of a practice of lending MSS.

The British Museum Aristotle.

[1] The list has been considerably increased by the recent discoveries of Messrs. Grenfell and Hunt at Oxyrhynchus, which include several literary fragments belonging to this period.

[2] Complete facsimile published by the Trustees of the British Museum (1891); specimen plate in *Pal. Soc.* ii. 122.

to be transcribed privately by those who wished to possess copies of the work; and if new papyrus could not be readily obtained. a roll which had already been used might be made to do duty again. This is what happened in the case of the Aristotelian treatise. It is written on the back of some old farm accounts; and the greater part of it is written in hands of a very cursive character. Apparently speed was necessary—perhaps the MS. from which it was copied could only be lent for a short time; for the different scribes evidently began independently, and their several sections do not exactly join. Two of the writers seem to be men of culture, able to write Greek correctly; and since their hands are similar in appearance (though certainly not identical, as has been suggested), it is perhaps not very fanciful to suppose that they are those of the owner of the MS. and a near kinsman. The other scribes seem to have been slaves, and their Greek is far from correct. One of them writes a sort of rough uncial hand, square, thick, and inelegant; the other, who takes up his work when he has finished it, tries at first to imitate his uncials, but quickly slides into a loose and ill-formed cursive hand, in which he completes his portion. Some of the more glaring blunders of the illiterate scribes are corrected in the hands of their educated collaborators. The date of the whole MS. is limited on the one side by the fact that the farm accounts on the *recto* are dated in the years 78–9; while the close resemblance of the cursive writing of the Aristotle to that of several documents (since discovered) of the reign of Domitian [1] makes it highly probable that it was not written very far from the year 90.

The autographs of the New Testament.

A somewhat special interest attaches to this MS., not only on the ground of its contents, but also because the method of private transcription, of which it is an

---

[1] See above, Plate VI, for an example. Another (of A. D. 83–4) is given in the *Führer durch die Ausstellung der Papyrus Erzherzog Rainer*, pl. ix.

example, must have been very commonly employed in
the case of the Scriptures of the New Testament. The
Christians of the first century were not predominantly
an educated body, nor were they, as a rule, wealthy.
Hence, in many parts of the Greek world, they could
neither transcribe their Scriptures in the best literary
hands of the period, nor pay professional scribes to do
so for them: and they must consequently have often
fallen back on the method of amateur, or private, transcription. This would especially be the case in times of
persecution, when the trade of transcribing the Scriptures
could not have been carried on publicly even in the
communities where there were otherwise facilities for
such work. In the writing of the Aristotle, therefore,
one may see an example of the manner in which the
Christian Scriptures were often transmitted; and in these
circumstances one may find part of the explanation of the
early rise of a large number of various readings. No
doubt there were exceptions, and some of the autographs
of the New Testament books may have been well and
carefully written; but the manner of transmission just
described must have been employed on all the books
in some stage of their history, and its characteristics
should therefore be borne in mind as a datum in the
textual criticism of the New Testament.

This, however, is somewhat of a digression from the
immediate subject in hand, the development of the literary
form of writing on papyrus; and it is time to return
to the other MSS. which, as mentioned above, seem to
fall within the latter part of the first century or in the
course of the second. One of these, which ranks next
to the Aristotle in point of literary interest, is also
remarkable as standing rather apart from the general
line of development, and provides a salutary proof of the
error of assuming that only one type of hand can have
prevailed at any one period. This is the papyrus of

Herodas[1], containing the mimes of that otherwise practically unknown writer of the Alexandrian age. As has
been stated in an earlier chapter, it is the smallest papyrus-
roll extant in respect of height, measuring no more than
5 inches, and containing only 18 lines, on an average,
in each column. As an almost necessary consequence, it
is written in a small and compact hand, clear and easy
to read, but almost entirely devoid of ornamentation.
Indeed the matter-of-fact realism of the poems seems
to be reflected in the plain, unadorned appearance of
the writing. The forms of the letters, when examined
in detail, are unquestionably of the Roman period, but
the general appearance of the writing is so unlike that
of any other extant papyrus, that it is exceptionally
difficult to fix its date from palaeographical considerations.
In the *editio princeps* it was assigned to the second or
third century, but increased knowledge makes it almost
certain that this date is too late. The clearest proof
of this comes from the accuracy with which non-literary
hands can now generally be dated. In the course of the
MS. a line has been accidentally omitted, and has been
supplied in the upper margin of the papyrus in a cursive
hand; and it happens that this contains an η of the peculiar form (γ) which has been described in a previous chapter
as characteristic of the period from about A. D. 50 to 160.
It is, no doubt, impossible to say how long after the
transcription of the MS. this omission was made good.
It may be said that the probabilities are against an error
in so comparatively rare a poet, and in a MS. in private
hands, being corrected from any except the original MS.
from which it was copied; but such a probability does
not amount to an argument of much strength. It is,
however, clear that the third century, and even the last

[1] Brit. Mus. Pap. cxxxv. Complete facsimile published by the Trustees
of the British Museum (1892); text, with specimen facsimile, in *Classical
Texts*; facs. of one column in Plate XVIII.

ΟΥΤΩϹΕΠΙΛΟΞΔΙΚΥΝΝΙΤΗΙΕΤΕΓΗΙΚΟΥΡΓΙ
ΑΝΘΟΙΝΑΥΦΙΛΗΓΑΡΔΙΕΦΕϹΙΑΥΧΘΡΕϹ
ΕϹΠΑΝΤΑΠΕϹΛΕΩΓΡΑΜΜΑΤΟΥΑϹΦΙΚΙΝΟϹ
ΩΝΘΡΩΠΟϹΕΝΛΛΕΝΙΛΕΝΕΝΔΑΠΗΡΝΗΘΗ
ΑΛΛΩΙΕΠΙΝΟΥΝΓΕΝΟΙΤΟΚΔΙΘΕΩΝΥΔΥΙΝ
ΗΠΙΓΕΘΟΟΓΔΕΚΙΝΟΝΙΗ(Ε)ΠΔΕϹΚΕΙΝΟΥ
ΛΛΗΤΠΑΛΙΦΔΛΛΠϹΔΕΕΚΔΙΚΠΩΡΩΡΗΚΕΝ
ΠΟΔΟϹΚΤΕΛΛΛΤΕΚΓΝΟϹΕΝΠΝΔΦΓΩϹΟ|ΚΩΙ
ΚΔΥΛΙΝΩΙΓΥΝΑΙΚΕϹΕΝΤΕΛΕΩϹΙΓΤΔΓΔ
ΧΝΙΕϹΛΩΙΟΝΕΛΛΛΕΠΟΝΤΔΛΙΓΩΝΩΟΥΤΙϹ
ΠΙΡΟΔϹΓΟΤΟΝΤΠΩΓΚΟΝΠΓΙΕΡΟΥΓΝΥΛΕΙϹ
ΙΠΙΗΓΤΟΙΜΟΝΘΥΛΕΓΙΗΓΕΙΙϹ
ΚΛΛΟΙϹΓΕΙΓΡΟΙϹΤΗΙΔΕΚΙΤΙΝΕϹΤΩΝΔΕ
ΕΔΟΓΙΝΙΤΓΤΗΙΤΕΚΜΓΕΝΠΓΑϹϹΟΝ
ΙΗΙΠΠΟΜΟΝΩΔΕΤΑΥΤΙ..
ΙΗΠΙΨΩΛΛΕΠϹΤΕΧΥΝΠΗΠΓΟΛΗΙ
ΕΛΘΟΙΛΕΝΑΥΠΓΜΕΖΟΝΠΔΝΙΝΓΥΟΜ
ϹΥΝΑΝΑΡΔϹ|ΝΚΛΠΩΓΚϹΤΤΔΝΙΚΔΝΩϹ
ΤΕΛΕΥΟΛΛΕΛΛΝΕΟΤΟϹΚΕΥΟΡΙΟΝΘΩΥΝΟΙ

HERODAS. – 1ST OR 2ND CENT.

part of the second, is too late a date to assign, and that the MS. should rather be placed in the first century or the first half of the second century. The forms of the individual letters require little notice; A is of the rounded type; M is deeply curved (μ, μ); Ξ has the top stroke separate, but the middle and lower strokes united (Ξ), a variant which *may* indicate a relatively early date, but is so rare as to provide no secure basis of argument [1]; Y is very stiff and straight, usually with a very short tail. Throughout it is a plain representation of Roman semi-uncial, with less grace than usual, but quite without affectation or mannerism [2].

Another MS. which should probably be placed about the turn of the century is the British Museum papyrus of Isocrates Περὶ Εἰρήνης [3]. It is a large papyrus, mutilated at the beginning, and written in two distinct hands, the first of which is decidedly better than the second. The latter indeed verges closely on the non-literary type of hand, and is certainly not the work of a trained scribe. The first hand has several features in common with the large Hyperides MS. A is rounded; Y is curved, not forked, which favours a date in the first century; C falls forward to some extent, which shows that it cannot be early in that century. The other letters are not distinctive. The second hand is loose and straggling to an extent which is not found before the latter part of the first

The British Museum Isocrates.

---

[1] This form recurs in the Oxyrhynchus fragment of Demosthenes, mentioned above (p. 76).

[2] Prof. Blass has recently, in a sort of *obiter dictum*, assigned the Herodas to the Ptolemaic period (*Bacchylidis Carmina*, pp. vii, viii), on the ground of its frequent interchange of ι and ει, which he regards as characteristic of this period ; but (1) a study of the non-literary papyri of the first and second centuries shows that such iotacisms were extremely common then (and this evidence is especially applicable to a MS. which, like the Herodas, is evidently not the work of a highly-trained scribe) ; and (2) the forms of the letters are wholly of the Roman type.

[3] Brit. Mus. Pap. cxxxii. Two specimen facsimiles in *Classical Texts*, with collation.

century at earliest. Its A, C, and Y are substantially the
same as those just described: and it has a very cursive
form of E. On the whole, the end of the first century
or the first half of the second seem to be the most
probable dates for this MS., which, it may be remarked,
is of some importance in the history of the transmission
of Greek literary texts, since it shows that, of the two
well-marked families into which the vellum MSS. of
Isocrates are divided, neither is exclusively to be trusted,
and that the formation of these families does not go back
to classical, or even to early Christian, times. It also
tends to show that the text of Isocrates was *substantially*
the same in the first century of our era as that which
has been preserved for us in MSS. a thousand years
later. But this is to trespass outside the domain of
palaeography.

We come now to a large group of Homer MSS., including
three somewhat small fragments in the Louvre, three long
rolls in the British Museum, and a long and very interesting MS. in the Bodleian. It would be absurdly rash
to try to place these in a precise chronological sequence;
but it is worth while to describe them severally, and to
show on what grounds they are assigned to the period
now under consideration.

Some
Homeric
papyri :
Brit. Mus.
Pap.
cxxxvi.

    In one instance there is external evidence in the shape
of dated writing on the other side of the papyrus. This
is Brit. Mus. Pap. cxxxvi, a roll containing nearly the
whole of the fourth book of the *Iliad*, and fragments
of Book III [1]. It is written in a rough, uneducated hand,
which gives it an appearance of later date than seems
to be actually true. The *recto* of the papyrus is occupied
by accounts, and among these are dates which must belong
to the reign of Augustus; and since a papyrus of accounts

---

[1] Specimen facsimile in *Classical Texts*, Plate vii. The MS. is there
wrongly dated, the evidence derivable from the writing on the *recto* not
having been discovered. Attention was first called to it by Prof. Wilcken.

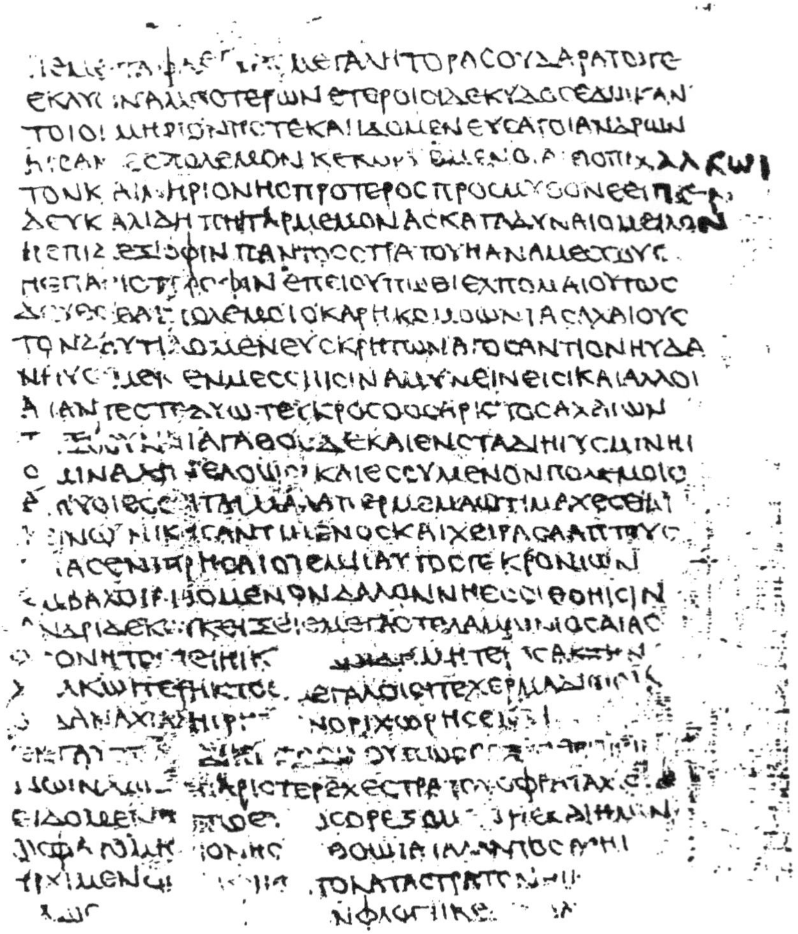

HOMER, ILIAD XIII.—1ST OR 2ND CENT.

would not be likely to be preserved very long (the case of the Aristotle MS. shows how soon such a papyrus might be used again), it seems almost certain that the Homer must have been inscribed upon the *verso* in the course of the first century. On the other hand the over-hanging shape of C and the v-shape of Y make it impossible to place it before the latter part of the century; and it may be even later than this. It is, however, so plainly not the work of a professional transcriber of literary MSS., that its importance in palaeographical history is not great.

Much more interesting, from the palaeographical point of view, is a papyrus brought from Egypt by Mr. Grenfell in 1896, and now in the British Museum (Pap. DCCXXXII)[1]. It is one of the longest Homer papyri in existence, containing the greater part of *Iliad* xiii and xiv, carefully written in a fine and ornamental hand. Indeed in point of appearance it is perhaps inferior to none but the British Museum *Odyssey* papyrus, described above. It has the uncial form of A—always a sign of a carefully written MS., but not confined to any one period; and the general aspect of the hand is square, firm, and well-formed. The other letters have the shapes which we have seen to be characteristic of the fully-developed Roman hand: E is well-rounded, with rather a long cross-bar; Z is square in form, and its cross-stroke not unfrequently meets the lower bar in the middle instead of at the end, which is an archaic feature; H has its cross-bar very high; the right-hand stroke of Λ projects but little above the point of junction; M and Ξ have their regular Roman forms; C is well rounded, but upright; Y is v-shaped; Ψ is very angular; and ω is rather stiff and compact. The whole writing is rather compressed, and has a tendency to lean backwards. So far as the forms just

Brit. Mus.
Pap.
DCCXXXII.

---

[1] A facsimile of a portion is given in Plate XIX. Alphabet in App. I, col. 15.

H

described differ from the average Roman hand of the
end of the first and the second century, they lean towards
the earlier forms; and consequently, if one were compelled
to assign a date within narrow limits, one would be disposed
to place it in the first century. It is, however, just a case
where additional care on the part of the scribe may make
a MS. appear older than it is, through his retaining the
older forms of letters; and it would consequently not be
wise to attempt to define the date too closely.

The same may be said of another Homer-papyrus, which,
in contrast to that just described, was the earliest of its
class to be discovered. This is the MS. known as the
Bankes Homer, from the name of the gentleman who
brought it from Egypt in the year 1821. It contains
the last book of the *Iliad*, in very good condition [1]. It
is written in a typically Roman hand, but with less
precision and compactness than the last example. A is
of the angular form which is seen as far back as the
first century B. C., but is more irregularly written; the cross-
bar of Є is low; Λ has a long projection above the point
of junction; Ϲ is round and upright; Υ is v-shaped and
rather straggling. The greater looseness and irregularity
of the writing suggest that it is somewhat later than
the last-mentioned MS., and that it should be placed in
the second century, as it has been by Sir E. Thompson;
but here again one does not know how much to allow
for the personal equation of the scribe. There are a few
corrections in a cursive hand which appears to be of
this century, but one cannot tell how long after the
original MS. they were written. An earlier date than
the first century, such as used formerly to be ascribed
to this MS., may be pronounced quite impossible in view
of modern knowledge.

The Homer papyri of the Louvre may be dismissed

The
Bankes
Homer.

---

[1] Now Brit. Mus. Pap. cxiv; specimen facsimile in *Cat. of Ancient MSS.*
(*Greek*), plate vi, and in *Pal. Soc.* i. 153. Alphabet in App. 1, col. 16.

more summarily, being very much smaller in extent and of no importance, either palaeographical or textual. Three fragments are included in the publication in the *Notices et Extraits* (nos. 3, 3bis, and 3ter)[1]. Of these the best from the palaeographical point of view is no. 3ter, a small fragment of what must once have been a handsome copy of the thirteenth book of the *Iliad*. It has the uncial A, but all the other test letters, such as M, Ξ and Y, are distinctly Roman. It may belong to either the first or the second century. Papyrus no. 3bis is less handsomely written, and has the rounded minuscule form of A; but there is nothing in the shapes of the letters to show that it is substantially later than the others. The remaining fragment, no. 3, is the least ornamental of the three, and probably the latest in date, being written in a somewhat loose and shaky hand. It does not appear, however, to be later than the second century, and shakiness of hand is less to be relied on as a sign of a relatively late date in papyri than in vellum MSS.; for the latter are almost invariably the work of professional scribes, and are generic rather than individual in their styles, whereas a large number of papyrus MSS. were written by amateur or semi-amateur transcribers, and reflect the idiosyncrasies of the individual as well as the tendencies or fashions of the age in which he lived. In cases such as this, therefore, we must be content with a rather wide range of uncertainty.

The first and second centuries represent the prime of the Roman style, and the principal MSS. belonging to them have now been described. It will have been seen that it is normally a hand of moderate size, with letters square in shape and well-rounded[2]; moreover, so far as we have yet

The Louvre Homers.

---

[1] Facsimiles (by hand) in the atlas accompanying vol. xviii. of *N. et E.*, plates xii and xlix.

[2] The terms are not contradictory, 'square' implying that the letters, whether angular or curved in outline, may be enclosed within a square.

gone, the letters are upright, even though the columns
of writing sometimes lean to the right.   It is a neat,
orderly, business-like hand, and in its best manifestations
graceful and, so far as is consistent with simplicity,
ornamental.   It is very probable that a fragment of the
Orestes of Euripides, now at Geneva, should be added to
the list of beautifully-written MSS. of this period; but
its editor[1], Prof. Nicole, who describes it as the handsomest
example of writing upon papyrus known to him, does not
fix its date precisely, and no facsimile of it has been
published.   Certainly, in the present writer's judgement,
the best examples of the Roman period excel those of any
other period in grace and beauty, as well as in general
workmanlike qualities; and this without in the least depre-
ciating the attractiveness of the strength and precision
seen in the best Ptolemaic papyri, such as the Louvre
Hyperides or the British Museum Bacchylides[2].

Ptolemaic letters are generally broader than they are high, and Byzantine
higher than they are broad, while Roman are normally about equal in
each dimension. And whereas Ptolemaic letters are angular, Roman have
their curves well rounded.

In the *Revue de Philologie*, xix. 105.

[2] The account which has been given of the palaeography of the first and
second centuries is confirmed by the recently-published Oxyrhynchus
papyri.   The published facsimiles from these papyri include a small but
beautifully-written fragment assigned to Alcman, apparently of the end
of the first century; a fragment of Thucydides (Book iv.) in a small,
irregular hand, probably of the latter part of the first century, but
recalling in its size the much earlier papyrus of the third Demosthenic
epistle p. 78 above); and the fragment of the Προοίμια Δημηγορικά of
Demosthenes, to which reference has been made above in connexion with
the Bacchylides MS., written in a broad, slightly sloping hand, rather
loose in formation.   Some other fragments of later date will be mentioned
below.   Another recently-published papyrus of this period is a mathe-
matical fragment known as the Ayer papyrus, brought from Egypt in
1895, and now in the Field Museum at Chicago (edited, with facsimile,
by Mr. E. J. Goodspeed in the *American Journal of Philology*, xix. 25).   It is
written in an ornamental hand approaching the uncial type, and has
been assigned by some authorities to the second or even the third century,
but would appear from the facsimile to be not later than the first.   Some
of the letters, especially Υ, are quite of the early Roman type.

PLATE XX.

HOMER, ILIAD II.—2ND CENT.

One MS. has, however, been reserved for separate con- sideration, which has a strong claim to be regarded as yet handsomer than any that has hitherto been mentioned. This is a copy of the second book of the *Iliad*, found at Hawara by Mr. Flinders Petrie, and now in the Bodleian Library; and it has been reserved on account of the doubt attaching to its date. Sir E. Thompson, whose judgement in palaeographical matters stands unquestionably first, originally assigned it to the fifth century, on account of its resemblance to the vellum MSS. of that date; and it is with great diffidence that a different opinion is here propounded. The MS. is written in an exceptionally large and carefully-formed hand[1]. The letters are square in build, upright in position, with well-rounded curves. A is of the uncial shape, with the cross-bar rather high; Є has the cross-bar rather low, as also has H; M is deeply curved; Y is Y-shaped, but not of the Ptolemaic or early Roman type. The other letters possess no distinctive characteristics, but are square, well-curved specimens of the best Roman hand. Now in many respects the characteristics here described apply to the vellum uncials of the fifth and sixth centuries, as they are seen in such MSS. as the Codex Alexandrinus and the Codex Claromontanus; and it is on this ground that Sir E. Thompson assigns to the papyrus a date contemporaneous with these MSS. On the other hand it will be seen that, except in point of size, the forms of the letters are just those of the Roman hand of the first and second centuries. At the time when the papyrus was discovered, no uncials of this size had been found among the papyri, and it was perfectly natural to associate them with the vellum uncials to which they unquestionably bear a considerable resemblance. The position has, however, since been altered by the discovery of such documents as the lease of A. D. 88,

---

[1] Facsimile of one column in Petrie's *Hawara, Biahmu, and Arsinoe*; part of a column in plate XX. Alphabet in App. 1, col. 17.

described above, and the fragment of the Λακεδαιμονίων
Πολιτεία, which show that a large uncial hand was in
use upon papyrus as early as the first or second century.
Hitherto the genesis of the hands seen in the early vellum
uncials has been unknown; but it now begins to seem
probable that they are the direct descendants of the hand
which had been used for the best papyrus MSS. two
hundred years earlier.   Further reasons in support of this
view will be found below, and will be considered in the
next chapter, in connexion with the transition from
papyrus to vellum : but meanwhile there is another argument in favour of an early date for the Oxford Homer.
A few notes have been added in the margins by a hand
different from that of the text; and these are written
in a sloping hand which will be seen presently to belong to
the third century [1].   No doubt a sloping hand also followed
the vellum uncials of the fifth and sixth centuries; but
it is to be remembered that after the beginning of the
seventh century Greek writing upon papyrus becomes very
much less common, and there is nothing Byzantine in
the hand in which these notes are written.

It is not maintained that these arguments are conclusive,
especially as the history of large uncial writing on papyrus
is not by any means fully known to us; but they seem
to establish a strong probability that the Homer may
be placed as early as the second century.   Some slight
confirmation of this conclusion may be found in a comparison of this MS. with the two minute scraps which
appear at the bottom of the plate containing the fragments
of the large Hyperides MS. (plate XVI).   These scraps,
coming from what must have been a very handsome copy
of Demosthenes' *Olynthiacs*, were found in a dummy
roll of papyrus along with the Hyperides fragments and

---

[1] One of these notes may be seen in the right-hand margin of the
facsimile ; but it is so faint in the plate as to be of little use to any one
who has not seen the original.

other material, none of which was later than the second century. There is therefore some slight presumption that they were themselves not very much later than that period; and it will be seen that there is a considerable resemblance between them and the Oxford Homer. It may be doubted also whether so handsome a copy would have been written on papyrus at a time when the substitution of vellum for that material had already been in progress for nearly a century; but this argument cannot be pressed, since such a question might be affected by the idiosyncrasies of an individual book-lover. The safest conclusion appears at present to be that the probabilities point to the second century as the date for this handsome MS., but that it would not be safe to treat it as an ascertained fact which may be used for the dating of other papyri. There can be very little doubt that the point will be settled before long by the discovery of fresh materials [1].

A short mention is perhaps due, in conclusion, to a MS. of considerable literary importance belonging to this period, though it falls rather outside the sequence of palaeographical development; namely the papyrus which contains the *Funeral Oration* of Hyperides [2]. Like the papyrus of Aristotle, the oration is written on the *verso* of the roll, and in a private and non-literary hand. It is not even, as the Aristotle is, a copy made by some

The Funeral Oration of Hyperides.

[1] An important piece of evidence in support of the conclusion stated above has already come to light in one of the Oxyrhynchus papyri (Grenfell and Hunt, *Oxyrhynchus Papyri*, i no. 20. plate v). This is, curiously enough, also a copy of the second book of the *Iliad*, and is written in a large and ornamental hand, strongly resembling that of the Oxford Homer; while its date is approximately fixed by the occurrence on the *verso* of some accounts which cannot be later than the third century. The Oxford Homer may, therefore, now be assigned to the second century with considerable confidence.

[2] Brit. Mus. Pap. xcviii; complete facsimile (by hand) in the *editio princeps* by Churchill Babington; specimen photographic facsimile in *Cat. of Anc. MSS. (Greek)*, plate iv.

private person desirous of possessing a work of classical antiquity. The character of the handwriting and the multitude of mistakes show that it must be a mere school-boy's exercise. Its literary value, in spite of the many blunders of transcription, is very great, since it alone preserves for us one of the most famous speeches of an orator who stood among the first in a time when the oratory of the world reached its highest pitch; but for the humbler purposes of palaeography it is not very useful. A schoolboy's exercise can throw little light on the writing of trained literary scribes, and holds no place in the history of palaeography. The date of the MS. is given approximately, though not precisely, by the writing on the *recto*, which consists of a horoscope, the year of which is shown by astronomical calculations to be either A. D. 95 or A. D. 155, with a decided probability in favour of the former. A horoscope would probably lose its interest after the lapse of a generation or so, and the oration may consequently be dated with practical certainty in the course of the second century[1]. The horoscope itself is written in a neat and graceful hand, which approaches the literary type, and serves to confirm the description which has already been given of the Roman hand. The same may be said of another horoscope in the British Museum (Pap. cx), which is dated in the year 138[2].

The third century.

The character of the book-hand of the first and second centuries has now been established as fully as the extant materials allow, and all the important documents which

---

[1] A MS. which in some degree recalls the writing of the *Funeral Oration* is published by Messrs. Grenfell and Hunt among the Oxyrhynchus papyri (no. 9, plate iii. It is a fragment of a treatise upon metre, probably by Aristoxenus, written in a rather ugly hand of medium size, sloping backwards, and in dark ink. It may be assigned to the latter part of the second or to the third century.

[2] Facsimiles of both these horoscopes are given in the atlas accompanying vol. i. of *Greek Papyri in the British Museum*.

come within that period have been described. In passing
to the third century we enter a region much more obscure.
There are no MSS. which can be assigned to it with
absolute certainty, or even with such a clear approach
to certainty as was the case with many of the papyri
which have hitherto been discussed; and there are not
many which can be placed in it on reasonably probable
evidence. There is, however, a group of MSS., exhibiting
the same general type of hand, which, in the light of
recent discoveries, appears to belong to this century; and
it is a group of some importance in palaeographical history.
The evidence on which this date is assigned to it is cir-
cumstantial rather than direct; but it seems to be fairly
conclusive.

The MS. on which the decision of the question mainly
turns is a papyrus in the British Museum containing the
second, third, and fourth books of the *Iliad* [1]. Unlike
all the papyri which have hitherto been described, it is
in book form, consisting of eighteen leaves, the text being
written on one side only. The most noticeable characteristic of the hand in which it is written is that it has
a decided slope. This in itself marks it off from the
upright hands of the first and second centuries, and it
also produces a lateral elongation of many of the letters
which is quite alien to the square build of the earlier
characters. A becomes broad, with a tendency to revert
to the angular shape of the later Ptolemaic style; M shows
a similar tendency, becoming at once wider and shallower
than before; Є is less well rounded; O is very small; Y is
angular and Y-shaped. The general aspect of the hand
is rough and coarse, but this may be due to the scribe,
who certainly was not a well-educated man, rather than
to the times in which he lived.

Brit. Mus. Pap. cxxvi.

---

[1] Pap. cxxvi ; specimen facsimile in *Classical Texts*, plate vi. The catalogue of ships is omitted from Book ii. and only the first hundred lines of
Book iv. are preserved. Alphabet in App. I, col. 18.

There is no direct evidence of the date of the MS., but some indirect evidence is provided by the writing which stands on the *verso* of the last four leaves, containing a short grammatical treatise attributed to Tryphon[1]. It is written in a fairly good cursive hand, rather small and compact, and leaning somewhat to the right. Until quite recently there was nothing to fix the date of this hand, so that it was of no use in establishing the age of the writing on the *recto*; but the clue has been furnished by some of the papyri brought home in 1896 by Mr. Grenfell. Some of these are dated in the reign of Diocletian, in the years just on either side of A.D. 300; and a comparison shows great similarity between these hands and that of the Tryphon. If, then, the Tryphon is to be placed somewhere about the year 300, the Homer must have been written earlier, and may safely be assigned to the third century. These dates are earlier than those which have previously been attached to the MS.[2]; but whereas the former opinion rested upon little but conjecture, the new dates are based upon fairly good evidence.

The results of this discovery are of some importance. In the first place it seems to carry with it the dates of a certain number of other documents, to which no dates could be satisfactorily assigned before, but which exhibit a sloping hand akin to that which has been described. For instance, there is a fragment of the well-known The Berlin theological work, the *Shepherd* of Hermas, published Hermas. (though not identified) by Wilcken[3]. The editor of this fragment, who stands quite at the head of the continental students of papyri, could not, at the time of the publication (1890), make any closer approximation to its date

---

[1] Specimen facsimile in *Classical Texts*, plate ix.

[2] In *Classical Texts* the Homer is provisionally assigned to the fourth or fifth century, the Tryphon to the fifth or perhaps the sixth.

[3] *Tafeln zur älteren griechischen Paläographie*, plate iii.

than to say that it must be some centuries older than a certain vellum MS. of the eighth century (the Fragmentum Mathematicum Bobiense) to which it bears some general resemblance. It is now, however, fairly clear that it must be placed in or near the third century, being written in a sloping hand akin to that of the British Museum Homer, though somewhat finer and thinner in style. The letters are of much the same shape, except that the curve of M is rather less shallow, and the hand as a whole shows the same evidence of looseness and degeneration.

Another fragment of some interest is the small piece of Ezekiel discovered by Mr. Grenfell, and now in the Bodleian, containing a Hexaplar text, with a few of the critical symbols employed by Origen [1]. Mr. Grenfell's second volume contains three additional examples of this type of hand, which may all be assigned to the same century; namely, the very interesting scrap of Pherecydes, a portion of a drama (there identified with the *Melanippe Desmotis* of Euripides, but perhaps a comedy), and parts of two columns of a prose work, which has been identified by Prof. Blass and others as a portion of the *Memorabilia* of Xenophon [2]. In the case of the drama, the evidence of date is strengthened by its possessing some scholia in a cursive hand which cannot be later than the third century.

Other MSS. in sloping hand.

The existence may consequently be taken to be established of a sloping type of hand, which is characteristic of the third century; and this confirms the conclusion to which reference has already been made, that the history of writing upon papyrus anticipates that of writing upon vellum, so that similar phenomena on different materials do not indicate contemporaneousness in date, but, on the

---

[1] Grenfell, *Greek Papyri*, i. 5 (without facsimile).

[2] Grenfell and Hunt, *Greek Papyri*, ii. 11–13; facs. of the Pherecydes and the drama, in plates iv and iii respectively.

contrary, admit of a difference of some centuries. On vellum a sloping hand is first found about the eighth century : but no one would propose to place the papyri which exhibit a sloping hand so late as this, when Greek writing was practically extinct in Egypt. If, however, these papyri are put back to the fifth or sixth century, as has hitherto been done, the principle of conformity with the vellum MSS. is abandoned, and there is consequently no reason why we should not carry them yet further back to the third century, if, as is now the case, trustworthy evidence points to such a conclusion.

The Marseilles Isocrates.

It so happens that papyri of this type are, at present, not at all plentiful in number, and nearly all that exist are mere scraps. Besides the Homer MS. already described, only one literary document of any size can as yet be placed in this category, namely the Marseilles papyrus of Isocrates *In Nicoclem*[1]. Concerning the date of this MS., which is of some value to the textual tradition of Isocrates, very diverse opinions have been expressed : the first editor, Schoene, assigning it to the Ptolemaic age, while B. Keil, who followed Schoene in discussing the MS.. places it in the fifth century after Christ. The former date is manifestly impossible ; the latter is probably too late. The MS. is written in two hands, the first a light, sloping hand, with affinities to those which have just been described, while the second is solider and heavier, having no very close parallels among extant papyri. It is, of course, possible that this hand shows us the successor of the third-century hand, but the differences in the shapes of individual letters are not very great, and it is more likely that both hands belong to the third century, and only differ through the personal idiosyncrasies of different scribes[2].

[1] Two specimen facsimiles (showing the different hands) in *Mélanges Graux*, with article by E. Schoene.

[2] The Oxyrhynchus papyri contribute some additional examples of this

In addition to this sloping book-hand, a semi-literary hand of a very different appearance is found in use in the latter part of the century. It resembles the cursive hand of the same period, which has been described in chapter III (p. 46), and is, in fact, that cursive hand adapted for literary use. An example of it may be seen in a fragment of a Psalter in the British Museum [1], which may apparently stand with the Oxford Ezekiel, a page of St. Matthew among the Oxyrhynchus papyri, and a scrap of Isaiah in the Rainer collection at Vienna, as the earliest extant remains of the Greek Bible. The hand is upright and somewhat showy, with the letters K and O especially prominent; but as it does not appear to be a regularly-established book-hand, it need not be described at length. The Psalter in question was apparently used as a school-book, having the separate syllables marked off by a different hand, to which is probably also to be ascribed some writing on the *verso* in which the syllables are actually written apart. This writing on the *verso* is in a cursive hand which belongs to the third or fourth century, and consequently confirms the date which has been assigned to the Psalter.

A hand very similar to that of the Psalter, though rather smaller, is found in a long magical papyrus in

British Museum Psalter, Pap. ccxxx.

sloping third century hand. The most important of these from the literary point of view is the fragment which is attributed, with great probability, to Sappho (no. 7, plate ii), and which is palaeographically notable as containing the only written example of a digamma, with the exception of one in the Paris Alcman. Still more important palaeographically is a fragment of Plato's *Laws* (no. 23, plate vi), on the *verso* of which is a date in the year A. D. 295, and which consequently confirms the attribution of hands of this type to the third century at latest. A fragment of St. Matthew (no. 2, plate i), which is likewise assigned to the third century, is written in too rough and untrained a hand to be of much palaeographical value.

[1] Brit. Mus. Pap. ccxxx; a roughly printed facsimile in *Athenæum* for Sept. 8 1894. The arguments there adduced for believing the papyrus to have been in book form do not appear to be sound.

the British Museum [1]; the semi-literary character of the
contents being suited well enough by the semi-literary
style of the writing.

The consideration of these MSS. has brought us to the
end of the third century, and here, with the re-organization
of the Roman world by Diocletian, the Roman period of
palaeography comes to an end too. In the domain of non-
literary papyri we have seen that this change of government
is accompanied by a change in the current style of writing,
which makes the division into periods real, and not merely
arbitrary. In the department of literary palaeography
there is likewise a change, which comes to pass at about
the same date, but it is a change of a different character,
the nature of which will be considered in the following
chapter.

To sum up in brief the history of the period which has
just been described, it will be seen that it covers a space
of some 350 years, as against the 250 or 200 which are
all that the present limits of our knowledge give to the
Ptolemaic age. The main characteristic introduced into
Greek writing during this period is that of roundness,
which is accompanied by increased fluency of style.
Letters are formed with curves instead of angles, and they
are written, as much as possible, without raising the pen.
The half-century which lies before the beginning of the
Christian era concludes the period of transition which had
begun under the later Ptolemies, and by about A. D. 1 the
Roman style, with its rounded curves and square build
of letters, is well established. The first century is the
period of its perfection; but before the close of it signs
of decay appear in the greater slackness and looseness
of formation in many letters. During the second century
this tendency is increased, and the appearance of firmness
combined with grace, which marks the earlier Roman

---

[1] Brit. Mus. Pap. cxxi ; facs. in atlas accompanying vol. i. of *Greek Papyri in the British Museum.*

hand, disappears from all except the best MSS. In the third century uprightness is exchanged for a sloping hand, —always a sign of degeneracy in palaeography; and this is generally accompanied by looseness and irregularity in the individual letters. The Roman hand has plainly run its course, and there is room for the new departure which has now to be considered.

CHAPTER VI

THE TRANSITION TO VELLUM

No Byzantine period of literary papyri.

IN treating of the non-literary papyri it has been shown that the Roman period is followed by an equally long and prolific period, to which the name of Byzantine has been given. But in the sphere of literary papyri there is no Byzantine period. It is true that a certain number of literary texts upon papyrus have been, at various times, located between the fourth and the seventh centuries; but reason has been shown in the previous chapter for transferring some of these to much earlier dates, while those which remain are, as will be seen presently, few and unimportant. Indeed the student who is acquainted with the principal extant papyri will already have recognized that hardly any of them still remain to be dealt with. The explanation is not to be found in any scarcity of papyrus as a writing material. The masses of documents of the sixth and seventh centuries which have been found in Egypt are sufficient to disprove any such theory. The cause is different, and is to be found in the successful rivalry of another material, namely parchment or vellum.

Early use of vellum.

The beginnings of the use of vellum are to be found much earlier, and are recorded in the standard works on palaeography. Without discussing the precise value of the story which tells of its invention in Pergamum in the second century before Christ, or the reference by Cicero to the use of διφθέραι[1], it is clear from Martial's *Apophoreta* that vellum was considerably used as a writing material in the second century of our era. It seems certain, however, that at this time it was generally regarded as inferior to papyrus. It was used for note-books, or for rough copies of literary compositions; and if, as Martial shows, copies of classical authors, such as Homer, Virgil, Livy, and Ovid, were sometimes made on vellum, they were common and cheap reproductions, better to give away to clients and hangers-on than to keep as ornaments of a good library.

The earliest extant example of writing upon vellum is perhaps a specimen of this very class of book. It is a somewhat recent acquisition of the British Museum, consisting of two small leaves of vellum, with two columns to each page, containing part of Demosthenes' speech *De Falsa Legatione*[2]. This fragment was found in Egypt, and though the exact circumstances of its discovery are not known, it came to Europe among a large collection of papyri, the majority of which belonged to the Roman period. The hand in which it is written bears a decided resemblance to that of the Herodas papyrus. It is quite unlike any known vellum hand, and we are therefore justified in regarding it as belonging to the period to which its appearance seems to assign it, namely, about the second century. It may thus fall within the period at which Martial was writing, and it corresponds very well to what has been said of the character of the vellum MSS. mentioned by the poet. It is plainly not an elaborately written copy. There is nothing of the appearance of an *édition de luxe*, as in the Vatican or Sinaitic MSS. of the Bible, or the Palatine and Medicean

Fragment of Demosthenes on vellum.

[1] *Epp. ad Att.* xiii. 24. See Thompson, pp. 35-37, for a summary of the data on this subject.

[2] Brit. Mus. Add. MS. 34473; text and description in *Journal of Philology*, xxii. 247 (1894).

Virgils. It may well have been regarded as an inferior class of book to the best papyrus MSS. of the period.

Absence of literary papyri after third century.
For the present, this example stands alone, and there is every reason to suppose that to the end of the third century papyrus held its own, at any rate in Egypt, as the material on which literary works were written. But when we come to the fourth century the situation is changed. For non-literary purposes, as has been shown above, the Roman period is succeeded by a Byzantine period, equally long and equally prolific, lasting well into the seventh century. Literary palaeography, however, has no Byzantine period, so far as papyri are concerned. It cannot be due to accident that scarcely any literary works on papyrus have come down to us which can reasonably be assigned to dates between the fourth and seventh centuries. It is true that there have been periods which, until quite lately, were unrepresented among extant papyri; but the cases are not parallel. In those instances no papyri at all of the periods in question had been discovered; which merely meant that the spade of the explorer had not hit upon any site of that particular date. In the case of the Byzantine period, on the contrary, not hundreds merely, but thousands of papyri have been found, which now choke the libraries of Berlin and Vienna; yet among these scarcely one contains literary work, and the few that do are palpably not written by professional scribes for the public market. It is, therefore, a fair conclusion that papyrus was not at that date being used so freely for literary purposes as it had been during the Roman period.

Vellum codices in the fourth century.
This inference from the negative argument of the absence of literary papyri later than the third century is confirmed by positive evidence from the other side. The fourth century is the date to which our earliest extant vellum MSS. (the above-mentioned Demosthenes excepted) are assigned; and though the grounds for this

ascription are not quite so conclusive as one could wish, they are sufficient to justify a fair amount of confidence. The MSS. in question include the Vatican and Sinaitic Codices in Greek, the Codex Vercellensis of the Old Latin Bible, the Codices Palatinus, Romanus, Vaticanus, and Sangallensis of Virgil, and possibly the Bembine Terence. The period of transition is further marked by the well-known statement of Jerome, that Acacius and Euzoius (towards the end of the fourth century) caused a number of damaged volumes in the famous library of Pamphilus at Caesarea to be replaced by copies upon vellum. The language employed by him[1] implies that this was a deliberate and systematic attempt to renovate the library; and the adoption of vellum in place of papyrus indicates that the victory of the former material had been gained by this time. It is consequently certain that the introduction of vellum had begun at a considerably earlier date. Finally, the evidence is clinched by the statement of Eusebius that the emperor Constantine, after his acceptance of Christianity, ordered fifty copies of the Scriptures to be prepared on vellum for use in the churches of Constantinople. When an emperor, on so special an occasion, orders vellum copies for the churches of his capital, in the beautifying of which no expense had been spared, we may be confident that vellum was by that time regarded as the best material for literary purposes.

It does not follow that papyrus was at once and alto- A few papyrus gether extinguished by the victory of vellum, and there MSS. of remain a few MSS. to be mentioned which appear to fourth century belong to the fourth or later centuries. In no case, how- or later. ever, are they documents of much palaeographical importance. A very long magical papyrus at Paris[2], and

---

[1] 'Quam [bibliothecam] ex parte corruptam . . . . in membranis instaurare conati sunt' (*Ep.* 141).

[2] Specimen facs. in Omont's *Facsimilés des plus anciens Manuscrits Grecs . . . de la Bibliothèque Nationale*, plate i.

I 2

a less long one in the British Museum [1], have been
assigned to the fourth century, though there is no direct
evidence to fix their date. They are written in book
form, on rather narrow pages (as in the case of the
papyrus of *Iliad* ii–iv, which has been assigned above
to the third century), and the writing slopes somewhat;
but it is not quite of the same type as the sloping hands
of the third century. It is a rounder, thicker, harder
hand, and its appearance suggests the possibility of the
writer having been familiar with vellum MSS. Further,
the text is written on both sides of the leaf, unlike the
Homer papyrus mentioned above; and this again seems
to indicate the influence of vellum. On the whole, the
case is one on which the only safe plea is that of ignorance;
but such indications as there are seem to point to the
fourth century rather than the third. The contents of
both documents are strongly coloured by Gnosticism, but
so intermixed with other elements that it is not at all
likely that they were written during the flourishing time
of that philosophy; while mistakes are so frequent as
to show that they are somewhat unintelligent copies, and
not original works. So far, therefore, as the contents
are concerned, there is nothing to discredit the date that
has generally been assigned to these MSS.

Psalter
in British
Museum.
Of more interest is a papyrus Psalter in the British
Museum [2], as to the age of which exceedingly different views
have been held. It consists of thirty-two leaves, rather
short and square in shape, containing Psalms x–xxxiv,
written in an irregular sloping hand. Tischendorf, who
first published its text, regarded it as earlier than any
vellum MS. in existence; but there is no reason to suppose
that he had ever given much attention to the palaeography

[1] Pap. xlvi; facs. in Atlas accompanying vol. i. of *Greek Papyri in the British Museum*, plates xxxvi—xlix.

[2] Pap. xxxvii; specimen facs. in *Cat. of Anc. MSS. (Greek)*, plate xii; *Pal. Soc.* i. 38.

of papyri, nor was there then sufficient evidence extant
to enable him to form an opinion. The editors of the
Palaeographical Society at first left it as early as the
fourth or fifth century, but subsequently brought it
down to the sixth or seventh; and there is little doubt
that the later date is correct. The hand is not that of
a trained scribe at all, and is not of the literary type.
It is not in the least like the sloping hand which we
have assigned to the third century. It is closely akin
to the non-literary hands of the late Byzantine period,
described in Chapter III. It will be remembered that
a sloping hand comes into use for non-literary documents
towards the end of the sixth century, and continues
into the seventh; and it is to this class of hands that
the Psalter in question belongs. It is of some importance
to fix the date, on account of the interest that would
attach to the MS. if Tischendorf's claim could be made
good; but its palaeographical importance is not great.

So far, the late papyrus hands have had no direct
and close connexion with the contemporary vellum hands.
There are, however, some cases in which there is a considerable resemblance. The earliest, in all probability, of
such MSS. is a fragment of Hesiod in the British
Museum [1], written in a large uncial hand of the same
general type as the Codex Alexandrinus. It is an upright hand, square in formation and well-rounded, but
differing from the second century MSS., of which the
same description has been given, in the thickness and
coarseness of the writing. This does not appear to be
due to a lack of skill on the part of the scribe, but to
be the result of deliberate intention. It is rather an
imitation of the earlier hand at a time when the light-
ness of touch, so necessary for good writing upon papyrus,
had been lost; and consequently there seems to be no

British Museum Hesiod.

---

[1] Pap. CLIX; text in *Revue de Philologie*, xvi. 181.

reason for separating it from the vellum hands to which it appears, *prima facie*, to be akin.

The Dublin St. Cyril.

Another papyrus, reproducing even more evidently the writing of contemporary vellum MSS., is a MS. at Dublin which contains some portions of the treatise *De Adoratione* of St. Cyril[1]. It is written in page-form in a good-sized uncial hand of the type that is generally called Coptic. It is especially marked by the upright A and the deeply curved M (Ϻ), which are found in a few vellum MSS., such as the Codex Zacynthius and the Codex Marchalianus. These MSS. are generally assigned to the sixth or seventh century, and the papyrus of Cyril may reasonably be placed in the same period. In a very similar hand is the fragment of a Festal Letter from a patriarch of Alexandria to his clergy, which was obtained by Mr. Grenfell in 1895–6, and of which the date is probably A.D. 577[2].

Papyrus of the Minor Prophets.

The last MS. that need be mentioned in this connexion is one of the prophets Zechariah and Malachi, which was formerly (and probably is still) in the possession of a Viennese dealer, Theodor Graf. It was exhibited at the London Oriental Congress of 1893, with the statement that its date was in the second century[3]. This date is, however, wholly impossible. The MS. is really of the same type as the last, but larger, rougher, and more irregular. It is a large, coarse uncial hand, written on pages of a good size. The leaves are bound together into quires by pieces of string. The seventh century is probably quite as early a date as it is entitled to claim, and, like the Cyril MS., it falls rather outside the domain of papyrus palaeography. The material on which it is written is indeed papyrus, but the writing is that which

---

[1] Published with facsimile in *Transactions of the Royal Irish Academy*, vol. xxix, pt. xviii, by the Rev. J. H. Bernard.

[2] Published with facsimile in Grenfell and Hunt's *Greek Papyri*, vol. ii ; now Pap. DCCXXIX in the British Museum.

[3] A rough facsimile was given in the *Times* during the Congress.

belongs properly to vellum. It shows a late stage in the transition from papyrus to vellum, when the victory of the latter had been won, though the Coptic Church continued for some centuries to use the ancient material for its vernacular Bible, and also, to some extent, for Greek copies of the Scriptures also.

Conversely, there are one or two MSS. which show upon vellum a style of writing to which we are accustomed only in papyri. The first of these, the fragment of Demosthenes' *De Falsa Legatione*, has been described already : the other which seems to deserve notice is the MS. of the Gospel and Revelation of Peter, discovered at Akhmim in 1886 and published in 1893 [1]. This has been assigned to the eighth century by some palaeographers, including the very high authority of M. Omont; but in truth there is no even approximate parallel to it among vellum MSS. Its kinsmen must be found in the Byzantine papyri of the sixth and seventh centuries, especially the former, as they have been described in Chapter III; and there seems to be no reason for declining to give this date to the Peter MS. It is confirmed by the other contents of the same volume, which are written in hands more closely conforming to the vellum type, and apparently of the sixth century. *(margin: Vellum MSS. in papyrus style.)*

Instances such as these, however, whether of vellum MSS. written in papyrus hands, or of papyrus MSS. in hands which belong properly to vellum, are so scarce as to be plainly exceptional, the natural concomitants of a period of transition. The general conclusion seems certain, that about the fourth century papyrus ceased to be the normal material for works of literature, and its place was taken by vellum. Of course this change of material involved a certain amount of change in the style of writing; but the general type remained the same, or rather, developed only in an ordinary manner and at an ordinary *(margin: The transition to vellum.)*

---

[1] Full facsimile by the French Mission Archéologique at Cairo; also by Gebhardt.

rate. The earliest vellum MSS. carry on the traditions of the Roman style of writing upon papyrus. The only change that is apparent is one in the direction of greater care and precision. The scribes found that the new material lent itself very easily to a handsome style of writing; and they consequently abandoned the somewhat slovenly sloping style which had come in during the third century, and modelled themselves on the earlier and more careful writing of the first and second centuries. The resemblance of the Sinaitic and Vatican codices to papyrus MSS. has been noticed more than once. Not only in the narrow columns, arranged four or three on a page, and showing eight or six at each opening of the book, but also in the hands themselves, there is a very real resemblance to papyri of the best Roman period. The Vatican MS. recalls such a papyrus as the British Museum *Odyssey*, or that which contains books xiii, xiv of the *Iliad*; the Sinaitic MS. reminds one still more strongly of the great Hyperides papyrus. The chief variation is in the increased thickness of the heavier strokes, which are differentiated from the light strokes far more than was natural on the comparatively delicate papyrus; though this tendency actually reacts upon the papyri written after the vellum period had set in, and accounts for the greater heaviness and thickness mentioned above as characteristic of such papyri.

The dates of the early vellum uncials.

The question is sometimes asked whether the evidence derivable from papyri does not modify our views as to the dates of the early vellum uncials. It has been shown in the preceding chapter that a hand of the same general type as that of the vellum MSS. is found on papyrus as early as the end of the first century. Is it not possible that the great MSS. of the Greek Bible are earlier than they have hitherto been supposed to be? It is impossible to argue the question fully here, since its decision rests largely upon details not properly palaeographical, such as the presence or absence of certain divisions of the text of the New Testament (the Eusebian canons, the Ammonian sections, the Euthalian divisions of the Acts); but it may be stated that the palaeographical evidence does not require any departure from the dates which have become traditional. In the case of the Codex Alexandrinus, the presence of a treatise of Athanasius (attached to the Psalms) precludes a date earlier than the latter part of the fourth century; while the Codex Sinaiticus has two notes or colophons which contain references to the library of Pamphilus at Caesarea—a library only formed at the beginning of the fourth century. On the other hand, the style of the beginning of the sixth century is fixed by the Vienna MS. of Dioscorides written for Juliana Anicia, daughter of Flavius Anicius Olybrius, who was emperor in the west in A. D. 472. A fuller discussion of this subject would belong rather to an essay on vellum-palaeography; but the facts here summarized show that the dates of the early Bible uncials do not rest upon palaeographical considerations alone, and are too firmly fixed to be affected by the evidence of the papyri. The same cannot be said of the very fine MS. of Homer, with illustrations, now at Milan[1]. The hand of this has a particularly early appearance; and though it would be rash to express a definite judgement merely on the strength of a few facsimiles, it is worth while suggesting a doubt whether this MS. may not be considerably older than the fifth century, the date to which it is now assigned.

The conclusion, therefore, to which the evidence leads is that the literary tradition passed wholly from papyrus to vellum in the fourth century, the exceptions in either direction being too few and too unimportant to require much notice. It appears, further, that the hands of the early vellum uncials are the direct descendants of the

---

[1] Specimen facsimiles in *Pal. Soc.* i. 39. 40, 50, 51.

hands found in literary papyri, especially those of the first and second centuries. The results of this revolution are not solely of palaeographical importance. They have also had a bearing on literary history, and have fundamentally affected our present knowledge of ancient literature.

Results of the change from papyrus to vellum. So long as papyrus was the material, and the roll the form, in common use for literary purposes, it was impossible to include works of very great extent in a single volume. Very large rolls were cumbersome to use, and perhaps were liable to be torn more easily than those which were lighter and shorter; in any case, as has been shown in an earlier chapter, a length of 20 to 30 feet is the greatest to which a single roll attains, while most of those which are now extant must have been considerably shorter. Consequently works of any great length could not be contained in a single roll, but had to be divided into a number of rolls. No papyrus roll of Homer hitherto discovered contains more than two books of the *Iliad*[1]. Three short orations fill the longest extant roll of Hyperides. The Ἀθηναίων Πολιτεία occupies no less than four rolls, though these, it is true, are of very small dimensions. A single play, a single long oration, would be enough to fill an ordinary roll. In the domain of sacred literature, the four Gospels and the Acts must each have occupied a separate roll; and no complete New Testament was possible until vellum came into use. It is obvious that conditions such as these favour *selection* among the works of an author. The ordinary literary student would not possess a complete Aeschylus or Sophocles, but only his favourite plays; and, since opinion tends to run in similar directions, some plays would be very much more in request than others. It is this fact that accounts for the state

---

[1] There is perhaps an exception to this rule in a papyrus recently discovered by Mr. Grenfell, and now in the Bodleian, the fragments of which contain lines from books xxii, xxiii, and xxiv. This MS. is in the small hand of the third century B. C., which would admit of a great amount of text in a roll of ordinary size.

in which the tragedians have come down to us. We have only a small proportion of their plays, but we have all that were most admired in antiquity. The meaning of this is that, at the time when vellum came into use and large collections became possible, the great majority of the dramas had already become scarce, and consequently few save those which we now possess were transcribed into vellum codices. We can imagine a similar state of things if the plays of Shakespeare had continued for two or three centuries, or even more, to circulate only in their original quarto form, where each play occupies a single volume, instead of being collected into the folios shortly after his death. In that case posterity would, no doubt, possess *Hamlet, Macbeth, As You Like It,* and *Much Ado about Nothing*; but it is highly probable that *Timon of Athens, All's Well that Ends Well, Pericles,* and *Titus Andronicus* would have been lost, or would survive only in fragments.

Other causes no doubt co-operated largely in determining the survival or disappearance of the various works of ancient literature. The stream of time, though far from acting as Bacon, in a pessimistic frame of mind, declared, has yet been singularly capricious in its choice of the cargoes which it has brought safely down to us. But among the conditions which have exercised an important influence upon its operations are, beyond doubt, the use of papyrus and of the roll-form as the vehicle of literary tradition for six centuries after the great Attic period, and its supersession by vellum, and by the modern book-shape or codex, in the fourth century of our era.

The influence of papyrus palaeography does not, however, end here. For non-literary purposes, for the ordinary occasions of daily life, papyrus continued to be used almost exclusively, at any rate in Egypt, and apparently elsewhere, for several centuries after its practical disappearance from the literary field. The development of

The minuscule hand on vellum developed from late papyrus hands.

cursive writing for private and business purposes runs
its regular course, as has been traced in the third chapter
of this essay, until the seventh, or even the eighth century.
Then our evidence ceases, through the accidental fact that
the use of Greek was about that time obliterated in the
only country from which materials for this department
of palaeography are derivable. But common sense tells
us that elsewhere in the Greek-speaking world—in Constantinople, in Athens, in Asia Minor—Greek cursive
writing continued to be used for every-day purposes,
whether upon papyrus or (as no doubt was increasingly
the case after the severance of Egypt from the Empire)
on vellum. The stream runs underground for a space,
but it reappears at the time of the palaeographic revolution
of the ninth century, when minuscules superseded uncials
as the current book-hand. As has been briefly stated by
Wilcken [1], in a sort of *obiter dictum*, the minuscule of the
tenth century is the descendant of the papyrus-cursive
of the seventh. The interval of time is so considerable
that very close resemblances are hardly to be expected,
especially since the tenth-century hand was a refined
and idealized reproduction of its archetype; but the
resemblance is quite traceable. The forms of letters in
the late Byzantine papyri are already approaching those
of the vellum minuscule; and when they are written
small, as in the receipts and accounts of the seventh
century, the resemblance is still more apparent. η has
developed its h-shape, and μ has its long perpendicular
tail; ν has its familiar v-shape, and υ has lost its tail
and become a simple curve; α, δ, ϵ, and indeed most
of the letters, approximate to the forms in which we
know them now. In the better examples there is even
something of the sharpness and precision which mark
the early vellum minuscules. They are not, it is true,
nearly so beautiful as the MSS. of the tenth century

_____

[1] In the Introduction to his *Tafeln*.

—the perfection of Greek minuscule writing—but they show plainly that they stand in the line of their ancestry.

It is so that, for the first time, the minuscule hand enters the line of literary tradition. The hand which for fifteen hundred years had been confined, so far as its proper use was concerned, to accounts, wills, and other such legal and private documents, and had only accidentally, as it were, become at times the vehicle of literature, now, in an improved and purified form, becomes the regular repository of literary works. The minuscule hand, even in its most cursive manifestations, is as old as the set uncial, so far at least as our records go back, and comes down side by side with it, though confined to less dignified functions. But at the last, when the uncial hand was exhausted, and could no longer supply the needs of increasing literary demands, the minuscule hand came forward to take its place, and to carry on the torch of literature for the five hundred years that still remained before the invention of the printing-press.

So, up to the end, the influence of papyrus remains traceable. Its immediate literary function was over when its book-hand gave birth to the vellum uncials of the fourth century; but its non-literary hand carried on a tradition which was to claim its place in literature six hundred years afterwards. Then its duties were at an end, and it is only in its remote descendants that it still survives, in the Greek types of to-day, which have been imitated or developed from the written hands of the Middle Ages.

APPENDIX I

THE following table gives the alphabets of the chief literary papyri which have been described in the preceding pages. They are arranged in what is believed to be their chronological order, though in the cases of MSS. of the first and second centuries this order must be regarded as only approximate. In the process of reproduction the size of the letters has been slightly reduced throughout. The following are the MSS. from which the alphabets are taken.

1. Literary fragments (Grenfell and Hunt, Gk. Pap. II. 1 = Brit. Mus. Pap. DCLXXXVIII). Early third century B. C.
2. Petrie *Phaedo* (Brit. Mus. Pap. CCCCLXXXVIII). Third century B. C.
3. Petrie *Antiope* (Brit. Mus. Pap. CCCCLXXXV). Third century B. C.
4. Dialectical Fragment (Louvre Pap. 2). Before 160 B. C.
5. Hyperides, *In Athenogenem* (in the Louvre). Second century B. C.
6. Bacchylides (Brit. Mus. Pap. DCCXXXIII). First century B. C.
7. Hyperides, *In Philippidem* (Brit. Mus. Pap. CXXXII). First century B. C.
8. Demosthenes, *Ep. iii.* (Brit. Mus. Pap. CXXXIII). First century B. C.
9. Herculaneum Pap. 152. First century B. C. Other Herculaneum MSS. have similar alphabets.
10. Homer, *Iliad* xxiii, xxiv. (Brit. Mus. Pap. CXXVIII). Late first century B. C.

11. Petition (Brit. Mus. Pap. cccliv).   *Circ.* 10 B. C.

12. Homer, *Odyssey* iii. (Brit. Mus. Pap. cclxxi).   *Circ.* A. D. I.

13. Homer, *Iliad* xviii. (Harris Homer = Brit. Mus. Pap. cvii).   First century.

14. Hyperides, *In Demosthenem*, etc. (Brit. Mus. Papp. cviii, cxv). Late first century.

15. Homer, *Iliad* xiii, xiv. (Brit. Mus. Pap. dccxxxii).   First century.

16. Homer, *Iliad* xxiv. (Bankes Homer = Brit. Mus. Pap. cxiv).   Second century.

17. Homer, *Iliad* ii. (Bodleian MS. Gr. class. a. 1 (P)). Second century.

18. Homer, *Iliad* ii.–iv. (Brit. Mus. Pap. cxxvi).   Third century.

For the non-literary papyri, an excellent table of alphabets will be found in Sir E. M. Thompson's *Handbook to Greek and Latin Palaeography*.

ALPHABETS OF LITERARY PAPYRI.

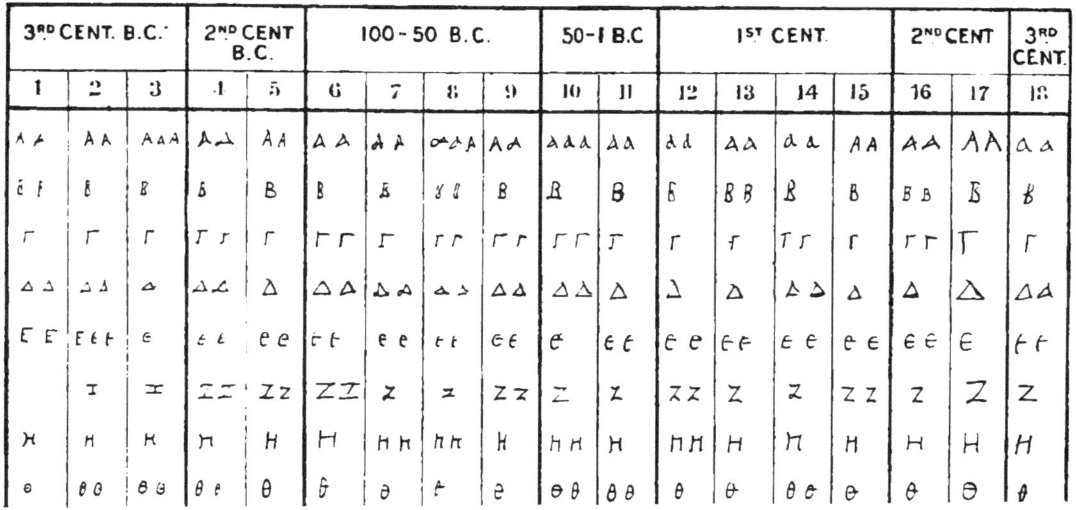
3RD CENT. B.C.
2ND CENT B.C.
100-50 B.C.
50-1 B.C
1ST CENT.
2ND CENT
3RD CENT.
1 2 3 4 5 6 7 8 9 10 11 12 13 14 15 16 17 18

APPENDIX II

THE following list was originally compiled independently, but since its preparation a similar list has been published by P. Couvreur in the *Revue de Philologie* (xx. 165), and another by C. Haeberlin in the *Centralblatt für Bibliothekswesen* (xiv. 1, 201, 263, 337, 389. 473), giving fuller details of the literature relating to each MS. From these lists some references have been taken; but the magical papyri (the principal publications of which are mentioned in App. III) and a few other non-literary works included by Haeberlin are not given here. The Oxyrhynchus papyri do not appear in either of the above-mentioned lists. In the case of papyri not in the British Museum, and of which no facsimile has been published, the dates given are those assigned by their editors. Where none is given, it will be understood that the editors have given none.

The following abbreviated references are used in this list:

Cat. of Additions = Catalogue of Additions to the Department of MSS., British Museum (published every six years).
Cat. of Anc. Greek MSS. = Catalogue of Ancient MSS. in the British Museum; part I, Greek. Edited by Thompson and Warner (1881).
Class. Texts = Classical Texts from Papyri in the British Museum; ed. Kenyon (1891).
F. or *Führer* = Führer durch die Ausstellung der Papyrus Erzherzog Rainer; (Greek section edited by Wessely (1894).
G. P. = Greek Papyri; vol. i by Grenfell (1896), vol. ii by Grenfell and Hunt (1897).
Mitth. or *Mitth. Erzh. Rainer* = Mittheilungen aus der Sammlung der Papyrus Erzherzog Rainer.

K

Notices et Extraits = Notices et Extraits des Manuscrits de la Bibliothèque
Impériale et autres Bibliothèques, vol. xviii (1865).

O. P. = Oxyrhynchus Papyri ; ed. Grenfell and Hunt (part i, 1898).

P. P. = Petrie Papyri ; ed. Mahaffy, two vols. (1891 and 1893).

Phil. Anzeig. = Philologischer Anzeiger (separate publication of Philologus)[1].

Rev. de Phil. = Revue de Philologie.

Rhein. Mus. = Rheinisches Museum.

Tafeln = Tafeln zur älteren griechischen Paläographie; ed. Wilcken (1891).

Wien. Stud. = Wiener Studien.

Zeitschr. f. äg. Sprache = Zeitschrift für ägyptische Sprache.

Aeschines, *In Ctesiphontem,* §§ 178–186. Fifth century.
Rainer Coll. Mentioned by Karabacek, *Mitth. Erzh. Rainer,*
i. 51.

Aeschylus, *Carians* (?) ; frag. (fifteen lines). Before B. C. 161.
Didot Papyrus. Weil, *Monuments Grecs* (1879), with facs.

— *Myrmidons* (?) ; frag. (eight lines). Before B.C. 161. Didot
Pap. Ed. Weil, with facs.

Alcidamas, Μουσεῖον ; fragment. Third century B. C. Brit.
Mus. Pap. ccccxcix. Mahaffy, *P. P.* i. 25, with facs.

Alcman, frag. of a lyric poem ; portions of three columns.
First century. Louvre Pap. 71. Egger, *Mémoires d'Hist.
anc. et de Philol.* 1863, p. 159 ; facs. in atlas to *Notices et
Extraits,* xviii., pl. 50.

Alcman (?), seven hexameter lines, three being imperfect, in
Aeolo-Doric dialect. First or second century. Grenfell
and Hunt, *O. P.* i. 8, with facs.

Apollonius, *Homeric Lexicon* ; fragment. Late first century.
Bodl. MS. Gr. class. e. 44 (P). Facs. by Nicholson.

Apollonius Rhodius ; fragment. Rainer Coll. Mentioned
in *Phil. Anzeig.* xv. 650.

---

[1] A number of literary fragments among the Rainer papyri are briefly
mentioned in various short notices in the *Philologischer Anzeiger* for 1884-6 ;
but it has seemed best not to include these, except when they are fairly
explicit. It is possible that mistakes were made in the first identifications,
and until fuller details are published these fragments cannot be considered
available for practical purposes.

Aratus, *Phaenomena* ; two fragments, in book form. Berl. Pap. Blass, *Zeitschr. f. äg. Sprache,* 1880.

Aristophanes, *Birds* 1057–1081, 1101–1127. Louvre Pap. Weil, *Rev. de Phil.* 1882.

Aristotle :—

Ἀθηναίων Πολιτεία. End of first century. Brit. Mus. Pap. cxxxi *verso.* Ed. Kenyon, 1891, with facs. ; subsequent editions by Ferrini, Kaibel and Wilamowitz, Herwerden and van Leeuwen, Blass, Sandys.

Ἀθηναίων Πολιτεία, two fragg., in book form. Fourth century (?). Berl. Pap. 5009. Blass, *Rhein. Mus.* 1880 ; Diels, *Abh. d. Berl. Akad.* 1885, with hand-made facs.

Post. Analytics, i. 71 B 19 — 72 A 38 ; in book form. Seventh century (?). Berl. Pap. 166. Landwehr, *Philologus,* xliv. 21 (1885).

Aristoxenus (?), ῥυθμικὰ στοιχεῖα, portions of five columns. Third century. Grenfell and Hunt, *O. P.* i. 9, with partial facs.

Bacchylides, *Odes* ; incomplete. First century B. C. Brit. Mus. Pap. DCCXXXIII. Ed. Kenyon (1897), with facs. ; subsequent editions by Blass and Jurenka (1898).

Basil, *Epp.* v. 77 E, vi. 79 B, ccxciii. 432 B, cl. 239 C, ii. 72 A. Fifth century (?). Berl. Pap. Blass, *Zeitschr. f. äg. Sprache,* 1880 ; Landwehr, *Philologus,* 1884 (with a facsimile).

Bible :—

Psalms xii. 7 — xv. 4. Late third century. Brit. Mus. Pap. ccxxx. *Athenæum,* Sept. 8, 1894, with facs.

— xi. 2 — xix. 6, xxi. 14 — xxxv. 6, in book form. Seventh century. Brit. Mus. Pap. xxxvii. Tischendorf, *Mon. Sac. Ined.,* nov. coll. i. 217 ; specimen facs. in *Cat. of Anc. Greek MSS.*

— xl. 16 — xli. 4, in book form. Berl. Pap. Blass, *Zeitschr. f. äg. Sprache,* 1881.

Bible (*continued*):—

Song of Solomon i. 6–9, in book form.   Seventh or eighth century. Bodl. MS. Gr. Bibl. g. 1 (P). Grenfell, *G. P.* i. 7.

Isaiah xxxviii. 3–5, 13–16. Third century. Rainer Pap. 8024 (*Führer*, 536).

Ezekiel v. 12 — vi. 3, in book form. Late third century. Hexaplar symbols. Bodl. MS. Gr. Bibl. d. 4 (P). Grenfell, *G. P.* i. 5.

Zechariah iv — *Malachi*?, in book form. Seventh century (?). Graf papyrus. Hechler, *Times*, Sept. 7, 1892, with specimen facs.

Matthew i. 1–9, 12, 14–20. One leaf of a book. Third century. Grenfell and Hunt, *O. P.* i. 2, with facsimile.

— xv. 12–16, fragment, Greek and Coptic, in book form. Sixth century. Rainer Coll. Mentioned by Gregory in Tischendorf's *Novum Testamentum Graece* iii. 450 (1884).

— xviii, fragment, in book form (?). Rainer Coll. Fourth or fifth century. Mentioned by Gregory, *ib.*

Mark xv. 29–38, in book form (? . Fourth century. Rainer Coll. Mentioned by Gregory, *ib.*

Luke v. 30 — vi. 4, in book form (attached to MS. of Philo, *vid. infr.*). Fourth century. Paris, Bibl. Nat. Scheil, *Mém. de la Mission Arch. Française au Caire*, tom. 9 (1893), with facs.

— vii. 36–43, x. 38–42, in book form. Sixth century. Rainer Pap. 8021 (F. 539).

John i. 29, in book form (?). Seventh century. Rainer Coll. Mentioned by Gregory, *ubi supra.*

1 *Corinthians* i. 17–20, vi. 13–18, vii. 3, 4, 10–14, imperfect, in book form. Fifth century (?). At Kiew, Uspensky Coll. Gregory, *op. cit.* iii. 344.

— i. 25–27, ii. 6–8, iii. 8–10, 20, in book form. Fifth century. At Sinai. Rendel Harris, *Biblical fragments from Mt. Sinai* (1890).

Bible (*continued*):—

Uncanonical Gospel; very small frag., parts of seven lines. Third century (?). Rainer Coll. Bickell, *Mitth. Erzh. Rainer*, i. 52 (1887), with facs.

Logia Iesu; one leaf of a book. Third century. Egypt Exploration Fund. Ed. Grenfell and Hunt (1897), with facs.; also in *O. P.* i. 1.

Chronological Treatise; fragment, portions of six columns, covering B. C. 355-315. Third century. Grenfell and Hunt, *O. P.* i. 12.

Cyril of Alexandria, *De Adoratione*, p. 242 E — 250 D, with lacunas, 286 B, in book form. Sixth or seventh century. Dublin Pap. Bernard, *Royal Irish Acad.* xxix. pt. 18, with partial facs.

Demetrius; philosophical works among Herculaneum papyri. First century B. c. See Scott, *Fragmenta Herculanensia*.

Demosthenes :—

In Aristocratem, lexicon to ; fragment. Fifth century (?). Berl. Pap. Blass, *Hermes*, xvii. 148.

De Corona, p. 308, small fragment. Third century. Grenfell and Hunt, *O. P.* i. 25, with facs.

Epistle III, wanting end. First century B. c. Brit. Mus. Pap. cxxxiii. Kenyon, *Class. Texts*, p. 56, with specimen facs.

De Falsa Legatione, § 10, imperfect. First or second century. Bodl. MS. Gr. class. f. 46 (P). Grenfell and Hunt, *G. P.* ii. 9.

Contra Leptinem, §§ 84-91, with lacunas. First or second century. Berl. Pap. 5879. Wilcken, *Tafeln*, i.

In Meidiam, §§ 41, 42. Fourth or fifth century. In possession of Mr. F. Cope Whitehouse. *Proceedings of Society of Biblical Archaeology*, xv. 86, with facs.

In Meidiam : hypothesis and part of commentary. Late first century. Brit. Mus. Pap. cxxxi *verso*. Kenyon, ᾽Αθηναίων Πολιτεία ed. 3, appendix i, with facs.

In Meidiam, lexicon to : fragment. Rainer Coll. Mentioned by Karabacek, *Mitth. Erzh. Rainer*, i. 51 (1886).

Demosthenes (*continued*):—

Olynthiacs II, fragments from §§ 10, 15. First-second century. In library of Rossall School. Kenyon, *Class. Rev.* vi. 430 ; facs. above, pl. XVI.

Third Philippic, fragment. Egypt Exploration Fund. Hogarth and Grenfell, *E. E. F. Archaeological Report*, 1895–6, p. 17.

Contra Phormionem, §§ 5–7, imperfect. Second century. Bodl. MS. Gr. class. f. 47 (P). Grenfell and Hunt, *G. P.* ii. 10.

Προοίμια Δημηγορικά, §§ 26–29, parts of seven columns. First or second century. Grenfell and Hunt, *O. P.* i. 26, with facs.

Dialectical Treatise, with quotations from Homer, Sappho, Alcman, Anacreon, Ibycus, Thespis, Timotheus, Euripides, &c. : fourteen columns, imperfect. Before B. C. 160. Louvre Pap. 2. Letronne, *Notices et Extraits*, xviii. 77, with facs. in atlas, pl. xi.

Dioscorides ; ten fragments in a chemical papyrus. Leyden Pap. x. Leemans, *Cat. of Leyden Papyri*, ii. 205.

Drama :—

Anonymous : fragment, apparently of an ' Iphigenia,' beginnings of seventeen lines. Third century B. C. Brit. Mus. Pap. ccccLXXXVI *b.* Mahaffy, *P. P.* i. 3 (2), with facs.

— parts of eighteen lines, apparently comedy. Third century B. C. Brit. Mus. Pap. ccccLXXXVII *a*. *Ib.* i. 4 (1), with facs.

— parts of twenty-eight lines, apparently tragedy. Third century B. C. Brit. Mus. Pap. ccccLXXXVII *b.* *Ib.* i. 4 (2), with facs.

— parts of twenty lines. Third century B. C. Brit. Mus. Pap. DXC. Mahaffy, *P. P.* ii. 49 *c.*

— two small fragments. Third century B. C. Brit. Mus. Pap. DXCI *a.* *Ib.* ii. 49 *d.*

— two small fragments. Third century B. C. Brit. Mus. Pap. DCLXXXVIII *a.* Grenfell and Hunt, *G. P.* ii. 1 *a*, with facs.

Drama (*continued*) :—

Anonymous : very small fragment. Third century B. C. Brit. Mus. Pap. DCLXXXVIII *b*. *Ib.* ii. 1 *b*, with facs.

— four fragments. Third century B.C. Brit. Mus. Pap. DCXC. *Ib.* ii. 6 *a*.

— fragment. Third century B. C. Brit. Mus. Pap. DCXCI *a*. *Ib.* ii. 6 *b*.

— two small fragments. Third century B. C. Brit. Mus. Pap. DCXCI *b*. *Ib.* ii. 6 *c*, with facs.

— two small fragments. Third century B. C. Brit. Mus. Pap. DCXCIV. *Ib.* ii. 8 *b*.

— fragment, fourteen lines. Before B. C. 161. Pap. Didot. Weil, *Monuments Grecs*, with facs.

— fragment of comedy, parts of fifty lines. First or second century. Grenfell and Hunt, *O. P.* i. 11.

— fragment. Second century. Brit. Mus. Pap. CCCCLXXXIV *d*. Unpublished.

— fragment of comedy, twenty lines, nine nearly perfect. Second or third century. Grenfell and Hunt, *O. P.* i. 10.

— parts of three or four columns, on subject of Jason. Second or third century. Brit. Mus. Pap. CLXXXVI *verso*. *Cat. of Additions*, 1894.

— fragment. Late third century. Brit. Mus. Pap. DCXCV *a*. Grenfell and Hunt, *G. P.* ii. 12, with facs.

— fragment, ten lines. Berl. Pap. Blass, *Zeitschr. f. äg. Sprache*, 1881.

Elegiac Poetry :—

Anonymous : fragment, parts of twenty-four lines. Third century B. C. Brit. Mus. Pap. DLXXXIX. Mahaffy, *P. P.* ii. 49 *a*.

— fragment, parts of eighteen lines. Second century. Grenfell and Hunt, *O. P.* i. 14.

Epic Poetry :—

Anonymous : two small fragments. Third century B. C. Bodl. MS. Gr. class. f. 45 (P). Grenfell and Hunt, *G. P.* ii. 5.

Epic Poetry (*continued*):—

Anonymous: fragment. Second century. Brit. Mus. Pap. cccclxxxiv *e*. Unpublished.

— considerable fragments. Second-third century. Brit. Mus. Pap. cclxxiii. *Cat. of Additions*, 1894.

— fragment, fifty-four lines. Fourth century (?). At Limerick (?). Bp. Graves, *Hermathena*, 1885, with facs.

— quotation, four lines. Fourth-fifth century. Paris, Bibl. Nat: Wilcken, *Sitzungsberichte der Berl. Akad.*, 1887.

— two fragments, on subject of Phineus. Rainer Coll. Mentioned in *Phil. Anzeig.* xiv. 477.

— fragment, in book form. Fourth century. Berl. Pap. 5003. Stern, *Zeitschr. f. äg. Sprache*, 1881; facs. in Wilcken, *Tafeln*, v.

Epicharmus; fragment, parts of four lines, from an anthology. Third century b. c. Brit. Mus. Pap. cccclxxxvi *a*. Mahaffy, *P. P.* i. 3 (1), with facs.

— Ὀδυσσεὺς αὐτόμολος, fragment, ten lines with commentary. First or second century. Rainer Pap. 8023 (F. 537, with facs.).

Epicurus, various works among Herculaneum papyri. First century b. c. *Cf.* Scott, *Fragmenta Herculanensia*.

Epigram, anonymous, on conquest of Egypt by Augustus. Early first century. Brit. Mus. Pap. cclvi (2). *Cat. of Additions*, 1894; *Rev. de Phil.* xix. 177.

Epigrams; collection by unknown authors, fragment. Third century b. c. Bodl. MS. Gr. class. e. 33 (P). Mahaffy, *P. P.* ii. 49 *b*, with facs.

— fragment of a collection. Third century. Grenfell and Hunt, *O. P.* i. 15.

Euclid, ii. 5. Third or fourth century. Grenfell and Hunt. *O. P.* i. 29.

Eudoxus, astronomical treatise. Early second century b. c. Louvre Pap. i. Brunet de Presle, *Notices et Extraits*, xviii. 25, with facs. in atlas, pl. i-x.

Euripides :—

Antiope, fragment from end of play, 123 lines. Third century B. C. Brit. Mus. Pap. ccccLxxxv. Mahaffy, *P. P.* i. 1, with facs.

— parts of three lines from an anthology. Third century B. c. Brit. Mus. Pap. ccccLxxxvi *a*. Mahaffy, *P. P.* i. 3 (1), with facs.

Medea, ll. 5–12. Before B.C. 161. Pap. Didot. Weil, *Monuments Grecs*, with facs.

Orestes, ll. 339–343, with musical notes. *Circ.* A. D. 1. Rainer Pap. 8029 (F. 531, with facs.). Wessely, *Mitth.* v. 65.

— ll. 1062–1090. *Circ.* second century. Geneva Pap. Nicole, *Rev. de Phil.* xix. 105.

[**Euripides**], *Rhesus*, ll. 48–96, in book form. Fourth–fifth century. Paris, Bibl. Nat. Wilcken, *Sitzungsb. d. Berl. Akad.* 1887.

Euripides, *Temenides* (?), forty-four lines. Before B. C. 161. Pap. Didot. Weil, *Monuments Grecs*, with facs.

Grammarian, anonymous ; fragment. First century. Rainer Coll. Mentioned by Karabacek, *Mitth.* i. 51.

Gregory of Nyssa, *Life of Moses* ; extracts. Fifth century (?). Berl. Pap. Blass, *Zeitschr. f. äg. Sprache*, 1880.

Hermas, *Pastor*, Mand. xi. 9, with an additional passage. Third–fourth century. Grenfell and Hunt, *O. P.* i. 5 (*cf.* V. Bartlet, *Athenæum*, Oct. 8, 1898).

— Sim. ii. 7–10, iv. 2–5. Third century. Berl. Pap. 5513. Wilcken, *Tafeln*, iii, Diels and Harnack, *Sitzungsb. d. Berl. Akad.* 1891.

Herodas, *Mimes* ; incomplete. First–second century. Brit. Mus. Pap. cxxxv. Kenyon, *Class. Texts*, with specimen facs. ; complete facs. in separate vol. Later editions by Rutherford, Bücheler, Crusius, Meister.

Herodotus :—

i. 76. Second or third century. Grenfell and Hunt, *O. P.*
i. 19 *verso*.

— 105-6. Third century. Grenfell and Hunt, *O. P.* i. 18.

Hesiod :—

Works and Days, ll. 111-118, 153-161, 174-182, 210-221,
with four additional verses; in book form. **Fifth century.**
Geneva Pap. Nicole, *Rev. de Phil.* xii. 113.

— ll. 251-266, 283-296, 313-329, 346-361, 686-709, 718-
740, 748-812, 817-828; *Shield of Heracles*, ll. 5-30,
434-440, 465-470; in book form. *Circ.* A. D. 400.
Rainer Coll. Wessely, *Mitth.* i. 73.

Theogonia, ll. 75-145, in book form. Fourth-fifth century.
Paris, Bibl. Nat. Wilcken, *Sitzungsb. d. Berl. Akad.*, 1887.

— ll. 210-238, 260-270. Fourth-fifth century. Brit. Mus.
Pap. CLIX. Kenyon, *Rev. de Phil.* xvi. 181.

Hesiod (?), *Eoeae* (?) ; portions of six lines. Third century
B. C. Brit. Mus. Pap. CCCCLXXXVI *c.* Mahaffy, *P. P.* i.
3 (3), with facs.

History :—

Anonymous, fragments of collection of νόμιμα βαρβαρικά.
Third century B. C. Brit. Mus. Pap. CCCCLXXXIX. Mahaffy,
P. P. i. 9, with facs.

— on Spartan training (? a Λακεδαιμονίων Πολιτεία) ; fragment.
Second century. Brit. Mus. Pap. CLXXXVII. Kenyon,
Rev. de Phil. xxi. 1.

Homer :—

Iliad i. 37-54, 65-67, 207-229 ; on the *verso* of accounts.
Third century (?). Brit. Mus. Pap. CXXIX. Kenyon,
Class. Texts, p. 80.

— i. 44-60. *Circ.* second century (?). Geneva Pap. Nicole,
Rev. de Phil. xviii. 103.

— i. 129-150. Second century. Brit. Mus. Pap. CCLXXII.
Cat. of Additions, 1894.

Homer (*continued*) :—

Iliad i. 273-362. Second century. Gizeh Museum (from Egypt Exploration Fund). Unpublished (*cf.* Hogarth and Grenfell, *E. E. F. Archaeological Report*, 1895-6, p. 16).

— i. 298-333. First or second century. Bodl. MS. Gr. class. c. 58 (f). Unpublished.

— i. 506, 507 ; ii. 1-6, 45-49, 111-115, 155-157, 200-205, 223-228, 245-252, 289-292, 331-337, 345-382, 391-404, 411-422, 433-446, 454-470, 472-486, 488-492, 494-510, 516-531, 538-560, 562-598, 601-621, 624-686, 692-731, 735-753, 755-841, 843-877 ; with few scholia and Aristarchean symbols. Second century. Bodl. MS. Gr. class. a. 1 (P). Petrie, *Hawara*, p. 24 ; specimen facs. there and above, pl. XX.

— fragments of *Iliad* i, ii, iv, viii, xi, xvii. Rainer Coll. Mentioned in *Phil. Anzeig.* xvi. 414, 477.

— ii. 101 — iv. 40 (omitting ii. 494-877), in book form. Third century. Brit. Mus. Pap. cxxvi. Kenyon, *Class. Texts*, p. 81, with specimen facs.

— ii. 730-828, with an additional line after l. 798. Second century. Grenfell and Hunt, *O. P.* i. 20, with partial facs.

— ii. 745-764. First or second century. Grenfell and Hunt, *O. P.* i. 21.

— iii. 317-337, 342-372 ; iv. 1-28, 56-69, 74-79, 111-150, 159-192, 198-201, 208-245, 256-293, 303-345, 352-544 ; on the *verso* of tax-register. Late first century. Brit. Mus. Pap. cxxxvi. Kenyon, *Class. Texts*, p. 93, with specimen facs.

— iv. 82-95. First-second century (?). Geneva Pap. Nicole, *Rev. de Phil.* xviii. 103.

— iv. 109-113. Third century b.c. Brit. Mus. Pap. DCLXXXIX *b*. Grenfell and Hunt, *G. P.* ii. 3, with facs.

— iv. 191-219, omitting 196, 197, 215. Gizeh Museum. Sayce, *Academy*, May 12, 1894.

— v. 731-734, 815-818, 846-850. Second century (?). Brit. Mus. Pap. cxxvii (1). Kenyon, *Class. Texts*, p. 98.

Homer (*continued*):—

Iliad vi. 1–39.   First-second century.   Louvre Pap. 3ᵗᵉʳ.
Longpérier, *Notices et Extraits*, xviii. 120, with facs. in
atlas, pl. xlix.

— vi. 90–100, 119–125.   Second century (?).   Brit. Mus.
Pap. cxxvii (2).   Kenyon, *Class. Texts*, p. 98.

— vi. 327–353 ; on *verso* of business document.   First-
second century (?).   Geneva Pap.   Nicole, *Rev. de Phil.*
xviii. 104.

— viii. 64–75, 96–116.   First or second century.   Bodl.
MS. Gr. class. d. 20 (P).   Grenfell, *G. P.* i. 2.

— viii, a fragment.   Second century B.C.   Egypt Exploration
Fund.   Hogarth and Grenfell, *E. E. F. Archaeological
Report*, 1895–6, p. 17.

— viii. 217–219 (?), 249–253, with two additional lines.
Third century B.C.   Brit. Mus. Pap. DCLXXXIX *a*.   Grenfell
and Hunt, *G. P.* ii. 2.

— viii. 433–447.   Berl. Mus. Pap. 6845.   *Verzeichniss der
ägypt. Altert.* p. 370.

— xi. 502–537, with five additional lines.   Third century B.C.
Brit. Mus. Pap. CCCCLXXXVI *d*.   Mahaffy, *P. P.* i. 3 (4),
with facs.

— xi. 788—xii. 9, with thirteen additional lines.   Second
century B.C.   Geneva Pap.   Nicole, *Rev. de Phil.* xviii.
104 : facs. in *Sitzungsb. d. Berl. Akad.* 1894.

— xii. 178–198.   Third century.   Bodl. MS. Gr. class. e.
21 (P).   Grenfell, *G. P.* i. 4.

— xiii. 26–47, 107–111, 149–173.   First or second century.
Louvre Pap. 3.   Brunet de Presle, *Notices et Extraits.*
xviii. 109, with facs. in atlas, pl. xii.

— xiii. 143–150.   Second century.   Vatican Pap. ; part of
same MS. as the preceding.   *Comptes-rendus de l'Acad.
des Inscr.* 1893.

Homer (*continued*) :—

Iliad xiii. 1–10, 38–50, 73–87, 149–437, 456–653, 658–674, 740–747 ; xiv. 120–293, 322–394, 397–522. First century. Brit. Mus. Pap. DCCXXXII. Hunt, *Journal of Philology*, xxvi. 25.

— xvii. 102–115, 142–152. Second century. Rainer Pap. 8027 (F. 533).

— xviii. 1–218, 311–617. First century. Brit. Mus. Pap. CVII. Thompson and Warner, *Cat. of Anc. MSS.* ; facs. in *Pal. Soc.* ii. 64.

— xviii. 1–22, 29–33, 77–92, 98–121, 125–136, 152–161, 168–175, 227–230, 273–275, 279–288, 320–349, 359–371, 387–394, 398–410, 412–425, 442–450, 455–465, 467–477, 479–492, 501–518, 534–543, 563–575, 578–617. Second century (?). Brit. Mus. Pap. CXXVII (3). Kenyon, *Class. Texts*, p. 98.

— xviii. 475–499, 518–535, 544–561. Second century. Louvre Pap. 3bis. Longpérier, *Notices et Extraits*, xviii. 114, with facs. in atlas, pl. xlix.

— xx. 36–110. Second century. Gizeh Museum (from Egypt Exploration Fund). Unpublished (*cf.* Hogarth and Grenfell. *E. E. F. Archaeological Report*, 1895–6, p. 17).

— xxi. 387–399, 607–611 ; xxii. 33–38, 48–55, 133–135, 151–155, 160–262, 340–344 ; xxiii. 159–166, 195–200, 224–229 ; with additional lines. Third century B. C. Bodl. MS. Gr. class. b. 3 (P). Grenfell and Hunt, *G. P.* ii. 4, with specimen facs.

— xxi. 544–609 ; xxii. 390–435. Berl. Pap. Blass, *Zeitschr. f. äg. Sprache*, 1880.

— xxiii. 1–79, 402–633, 638–897 ; xxiv. 1–83, 100–144, 150–243, 248–282, 337–341, 344–351, 382–387, 402–479, 490–520, 536–548, 559–579, 596–611, 631–657, 671–729, 737–743, 754–759 ; with a few scholia and Aristarchean symbols. First century B. C. Brit. Mus. Pap. CXXVIII. Kenyon, *Class. Texts*, p. 100, with specimen facs., and *Journal of Philology*, xxi. 296.

Homer (*continued*) :—

Iliad xxiv. 127–804. Second century. Brit. Mus. Pap. cxiv. Thompson and Warner, *Cat. of Anc. MSS.*, with facs.

Odyssey iii. 267–278, 283–294, 319–335, 352–366, 389–497, with scholia. Early first century. Brit. Mus. Pap. cclxxi, with some small fragments in Rainer Coll. Kenyon, *Journal of Philology*, xxii ; specimen facs. in *Pal. Soc.* ii. 182 ; Wessely, *Mitth.* vi. 1, with hand-made facs. of Vienna fragments.

— iii. 364–375, 384–402. Geneva Pap. Nicole, *Rev. de Phil.* xviii. 101.

— v. 346–353. Third century. Bodl. MS. Gr. class. g. 7 (P). Grenfell, *G. P.* i. 3.

— xiv. 15–24, 36–60, 71–86, 374–376, 378–381, 407–409, 430–441. Berl. Pap. 154 a. Landwehr, *Philologus*, xliv. 685.

— xv, fragment. Gizeh Museum. Sayce, *Academy*, May 12, 1894.

Homer : Lexicon to *Iliad* i ; fragment. Fifth century. Berl. Pap. Wilcken, *Sitzungsb. d. Berl. Akad.* 1887.

Homer : Commentary and paraphrase to *Iliad* i ; fragment. Third-fourth century. Paris, Bibl. Nat. Wilcken, *ib.*

Homer : Paraphrase of *Iliad* iv ; fragment. Rainer Pap. Mentioned in *Phil. Anzeig.* xiv. 44.

See also *s. v.* Apollonius.

Hyperides :—

In Athenogenem ; imperfect. Second century b. c. Louvre Pap. Ed. Revillout, with facs. ; subsequent editions by Weil, Herwerden, Blass, Kenyon, Vogt.

In Demosthenem, fragments ; *Pro Lycophrone*, imperfect ; *Pro Euxenippo*. Late first century. Brit. Mus. Papp. cviii, cxv (with fragments in Paris and at Rossall School). Ed. Babington, with hand-made facs. ; other editions by Sauppe, Boeckh, Schneidewin, Caesar, Cobet, Linder, Comparetti, Blass.

Hyperides (*continued*):—

Funeral Oration, imperfect. Second century. Brit. Mus. Pap. xcviii *verso*. Ed. Babington, with hand-made facs. ; subsequent editions by Sauppe. Kayser, Tell, Cobet. Comparetti, Dehèque, Caffiaux, Blass.

In Philippidem, imperfect. First century B. C. Brit. Mus. Pap. cxxxiv. Ed. Kenyon, *Class. Texts*, with specimen facs.: subsequent editions by Weil, Herwerden, Blass, Kenyon.

Isocrates:

De Antidosi, §§ 83, 87, small fragment. First or second century. Grenfell and Hunt, *O. P.* i. 27.

Contra Nicoclem, §§ 1–30. Third century. Marseilles Pap. Schoene, *Mélanges Graux*, 481, with specimen facs.

— §§ 2–4, fragment. Fourth century. Rainer Pap. 8029 (F. 532). Wessely, *Mitth.* iv. 136 (1888).

De Pace, §§ 1–61 fragmentary, rest nearly perfect. First–second century. Brit. Mus. Pap. cxxxii. Kenyon, *Class. Texts*, p. 63, with specimen facs.

Philippus, §§ 114–117. First–second century. Rainer Pap. Wessely, *Mitth.* ii. 74 (1887).

Isocrates: criticism of the *Evagoras*, anonymous. First–second century. Rainer Pap. Wessely, *Mitth.* ii. 79 (1887).

Literary Criticism:—

Anonymous, on the names of the gods ; fragment. Second–third century. Berl. Pap. 1970. Wilcken, *Tafeln*, ii.

— life of Aesop. One leaf of a book. Sixth century (?). At St. Petersburg. Weil, *Rev. de Phil.* ix. 19.

Lives of Saints, anonymous; lives of SS. Abraham and Theodora, fragments. Louvre Papp. 7404–8 bis. Wessely, *Wien. Stud.* 1889.

Lyric Poetry :—

Anonymous ; two small fragments. Third century B. C. Brit. Mus. Pap. dcxciii. Grenfell and Hunt, *G. P.* ii. 8 *a*.

Lyric Poetry (*continued*) :—

Anonymous ; fragment of ἐπινίκιον, twenty lines. Louvre
Pap. Egger, *Comptes-rendus de l'Acad. des Inscr.* 1877.

— portions of three columns. Second century. Egypt
Exploration Fund. Hogarth and Grenfell, *E. E. F.
Archaeological Report*, 1895–6, p. 16.

Mathematics, anonymous ; fragment on mensuration of land.
First century. Ayer Papyrus. E. J. Goodspeed, *American Journal of Philology*, xix. 25, with facs.

Medicine :—

Anonymous ; treatise on diseases, with extracts from Menon's
Iatrica. First–second century. Brit. Mus. Pap. cxxxvii.
Ed. Diels, *Supplementum Aristotelicum*, iii. 1, with speci-
men facs. (1893).

— fragment, on dentistry &c. First-second century. Brit.
Mus. Pap. clv. *Cat. of Additions*, 1894.

Menander, Γεωργός, portions of one leaf, about eighty lines.
Fourth-fifth century (?). Ed. Nicole (1897), Grenfell
and Hunt (1898).

Mime, anonymous. in rhythmical prose or lyrical verse ;
imperfect. Third century B. C. Brit. Mus. Pap. dcv
verso. Grenfell, *G. P.* i. 1, with facs.

Oratory : —

Anonymous ; fragment. First century. In Paris (?).
Egger, *Mém. d'Hist. anc.* 175 (1863).

— remains of three rhetorical exercises, mutilated. Late
first century. Brit. Mus. Pap. cclvi *verso*. *Cat. of
Additions*, 1894 ; Kenyon, *Mélanges Weil*. p. 243 (1898).

— two columns of a private oration. Second century.
Egypt Exploration Fund. Hogarth and Grenfell, *E. E. F.
Archaeological Report*, 1895–6, p. 16.

— fragments. In Paris (?). Late first or second century.
Egger, *Mém. de l'Acad. des Inscr.* xxvi (1870), with facs.

Pherecydes, Πεντέμυχος: fragment. Third century. Bodl.
MS. Gr. class. f. 48 (P). Grenfell and Hunt, *G. P.* ii. 11,
with facs.

Philo, τίς ὁ τῶν θείων κληρονόμος and περὶ γενέσεως Ἄβελ : in book form. Sixth century (? third). Gizeh Museum. Ed. Scheil, *Mém. de la Mission Arch. Française au Caire*, tom. 9 (1893), with specimen facs.

Philodemus, various works among Herculaneum papyri. *Cf.* Scott, *Fragmenta Herculanensia.*

Philosophy :—

Anonymous ; fragment. Third century B. C. Brit. Mus. Pap. DCXCII. Grenfell and Hunt, *G. P.* ii. 7 *a.*

— five small fragments. Third century B. C. Bodl. MS. Gr. class. e. 63 (P). *Ib.* 7 *b.*

— fragment. Third century B. C. Brit. Mus. Pap. DXCI *b.* Mahaffy, *P. P.* ii. 49 *c.*

— several rolls among Herculaneum papyri. First century B. C. *Cf.* Scott, *Fragmenta Herculanensia.*

— several fragments of a treatise, apparently philosophical, on *verso* of land-register. Second century. Brit. Mus. Pap. DCCXXXIV.

— life of the philosopher Secundus. Second century (?). At St. Petersburg. Tischendorf, *Notitia Editionis Codicis Sinaitici*, p. 69 (1860).

— fragment on ethics. Second or third century. Brit. Mus. Pap. CLXXXIV. *Cat. of Additions*, 1894.

— considerable fragments on ethics. Third century (?). Brit. Mus. Pap. CCLXXV. *Ibid.*

— (? Aristotle), fragment on aesthetics. Rainer Coll. Mentioned in *Phil. Anzeig.* xiv. 414; *cf.* Gomperz, *Mitth.* i. 84 (1887).

Plato :—

Gorgias, parts of pp. 504 B–E, 505 A, in book form. Third century. Rainer Coll. Wessely, *Mitth.* ii. 76 (1887).

Laches, small fragments of pp. 181 B–182 A. Second century. Brit. Mus. Pap. CLXXXVII *verso. Cat. of Additions*, 1894.

— pp. 190 B–192 A, with lacunas. Third century B. C. Bodl. MS. Gr. class. d. 22, 23 (P). Mahaffy, *P. P.* ii. 50, with facs.

Plato *(continued)* :—

> *Laws*, ix. pp. 862-3. Third century (before A. D. 295). Grenfell and Hunt, *O. P.* i. 23, with facs.

> *Phaedo*, pp. 67 E-69 A, 79 C-81 D, 82 A-84 B, with lacunas. Third century B. C. Brit. Mus. Pap. CCCCLXXXVIII. Mahaffy, *P. P.* i. 5-8, with facs.

> *Republic*, x. pp. 607-8. Third century. Grenfell and Hunt. *O. P.* i. 24.

Political Treatise, fragment of letter to a king of Macedon against the Thebans. Second or third century. Grenfell and Hunt, *O. P.* i. 13.

Posidippus, two epigrams. Before B. C. 161. Pap. Didot. Weil, *Monuments Grecs*, with facs.

Romance :—

> Anonymous, on adventures of Heracles ; fragments. Third century B. C. Brit. Mus. Pap. DXCII. Mahaffy, *P. P.* ii. 49 *f.*

> — on adventures of Ninus ; fragments. *Circ.* B. C. 50-A. D. 50. Berl. Pap. 6926. Wilcken, *Hermes*, xxviii. 161 (1893).

> — on Metiochus and Parthenope ; fragment. Second century. Berl. Pap. 7927. Krebs, *Hermes*, xxx. 144 (1895).

> — with narrative of shipwreck ; fragment, on *verso* of accounts. First-second century. In Dublin (?). Mahaffy, *Rendiconti della R. Accad. dei Lincei*, 1897, with facs.

> — considerable fragments. Second century. Brit. Mus. Pap. CCLXXIV. *Cat. of Additions,* 1894.

Sappho (?) ; five fragmentary stanzas. Third century. Restored by Blass, *ap.* Grenfell and Hunt, *O. P.* i. 7, with facs.

Scazon Iambics, anonymous ; fragment. Third century (?). Brit. Mus. Pap. CLV *verso*. *Cat. of Additions,* 1894.

Science :—

> Anonymous, on optics. Louvre Pap. 7733. Wessely, *Wien. Stud.* xiii. 312 (1891).

Science *(continued)* :—

Anonymous, chemical excerpts. Third-fourth century. Leyden Pap. x. Leemans, *Papyri Graeci Lugduni-Batavi.* ii. 199 (1885).

— on mathematics. Seventh century (?). Gizeh Museum. Ed. Baillet, in *Mém. de la Miss. Arch. Fr. au Caire.* tom. 9 (1892), with facs.

— on astronomy ; fragment. Rainer Pap. Mentioned in *Phil. Anzeig.* xiv. 477 (1884).

Sophocles, *Oed. Tyr.* 375–385, 429–441, in book form. Fifth century (?). Grenfell and Hunt, *O. P.* i. 22.

Theology :—

Anonymous ; fragment, sixteen lines. Sixth–seventh century. Brit. Mus. Pap. cxiii. 12 a. Kenyon, *Cat. of Papyri.* i. 224, with facs. (1893).

— fragment, thirty-two lines. Sixth–seventh century. Brit. Mus. Pap. cxiii. 12 b. *Ib.* p. 225, with facs.

— fragments, forty-six lines. Sixth–seventh century. Brit. Mus. Pap. cxiii. 12 c. *Ib.* p. 226, with facs.

— fragment, twenty-six lines. Sixth–seventh century. Brit. Mus. Pap. cxiii. 13 a. *Ib.* p. 227, with facs.

— fragment. Sixth or seventh century. Brit. Mus. Pap. cccclxii. Unpublished.

— fragment. Sixth or seventh century. Brit. Mus. Pap. cccclxiv. Unpublished.

— fragment. Berl. Pap. Blass, *Zeitschr. f. äg. Sprache.* 1881.

— fragment, on *verso* of papyrus. Fourth century. Grenfell and Hunt, *O. P.* i. 4.

Thucydides :—

ii. 7, 8. Second or third century. Grenfell and Hunt. *O. P.* i. 17.

iv. 36–41. First-second century. Hunt, *E. E. F. Archaeological Report,* 1896–7, p. 13, and *O. P.* i. 16, with partial facs.

Travels; fragment of a description of Athens. Second century. Petrie, *Hawara*, p. 28 (1889); *cf.* Wilcken, *Berl. Phil. Wochenschrift*, Dec. 7, 1889.

Tryphon, τέχνη γραμματική, imperfect, in book form. Early fourth century. Brit. Mus. Pap. cxxvi *verso*. Kenyon, *Class. Texts*, with specimen facs.

Xenophon :—

Cyropaedia, v. 2. 3–3. 26, imperfect. Second century. Rainer Coll. Wessely, *Mitth.* vi. 1 (1897).

Hellenica, i. 2. 2–5. 8, imperfect. Third century. Rainer Coll. Wessely, *ib.* vi. 17.

— iii. 1, parts of three columns. Second century (?). Grenfell and Hunt, *O. P.* i. 28.

Memorabilia, i. 3. 15, 4. 1–3. Third–fourth century. Brit. Mus. Pap. dcxcv *b*. Grenfell and Hunt, *G. P.* ii. 13.

APPENDIX III

(Arranged chronologically under the respective countries. Isolated publications in periodicals are not included ; they will mostly be found in *Hermes, Philologus*, and the *Revue de Philologie*.)

I. AUSTRIA.

1. PETTRETINI (G.), *Papiri greco-egizi ed altri greci monumenti dell' I. R. Museo di Corte* (Vienna, 1826), with three plates.
2. PEYRON (A.), *Papiri greco-egizi di Zoide dell' I. R. Museo di Vienna*, with two plates (Turin, 1828) ; a revision of some of Pettretini's texts.
3. WESSELY (K.), articles in *Wiener Studien*, iii–v, vii, xi (Vienna, 1881–1889).
4. WESSELY (K.), articles in *Mittheilungen aus der Sammlung der Papyrus Erzherzog Rainer* (Vienna, 1887–1897).
5. WESSELY (K.), *Papyrus Erzherzog Rainer: Führer durch die Ausstellung* (Vienna, 1894) ; edited by J. Karabacek, the Greek section by Wessely, with nine facsimiles of Greek papyri.
6. WESSELY (K.), *Corpus Papyrorum Raineri*, vol. i. Griechische Texte (Vienna, 1895).

II. FRANCE.

7. BRUNET DE PRESLE (W.), *Notices et Textes des Papyrus Grecs du Musée de Louvre et de la Bibliothèque Impériale* (in *Notices et Extraits des Manuscrits de la Bibliothèque Impériale et autres Bibliothèques*, vol. xviii), with atlas of fifty-two plates (1865).

8. WESSELY (K.), *Lettres à M. Revillout* (in *Revue Egyptologique,* 1884).

9. WESSELY (K.), article in *Wiener Studien,* viii (1886).

10. WESSELY (K.), *Griechische Zauberpapyrus von Paris und London* (*Denkschriften der kais. Akademie der Wissenschaften.* Vienna, 1888).

11. WESSELY (K.), *Die Pariser Papyri des Fundes von El-Faijûm* (*ib.* 1889).

12. WESSELY (K.), *Zu den griechischen Papyri des Louvre und der Bibliothèque Nationale* (in *Jahresbericht des k. k. Staatsgymnasiums Hernals,* two parts, Vienna, 1889, 1890).

13. WITKOWSKI (S.), *Prodromus grammaticae papyrorum Graecarum aetatis Lagidarum* (Cracow, 1897); includes revision of no. 7.

III. GREAT BRITAIN AND IRELAND.

14. FORSHALL (J.), *Greek Papyri in the British Museum* (London. 1839).

15. PEYRON (B.), *Papiri Greci del Museo Britannico di Londra e della Bibliotheca Vaticana* (Turin, 1841); Forshall's texts re-edited with commentary.

16. WESSELY (K.), articles in *Wiener Studien,* viii, ix, xii (Vienna, 1886, 1887, 1890).

17. WESSELY (K.), *Griechische Zauberpapyrus von Paris und London* (*Denkschriften der k. Akademie der Wissenschaften,* Vienna, 1888); = no. 10 above.

18. WESSELY (K.), *Neue Griechische Zauberpapyri* (*ib.* 1893).

19. MAHAFFY (J. P.), *The Flinders Petrie Papyri*; Part I, with thirty plates (Dublin, 1891); Part II, with eighteen plates (Dublin, 1893); appendix, with three plates (Dublin, 1894).

20. KENYON (F. G.), *Greek Papyri in the British Museum. Catalogue with Texts*; vol. i, with atlas of 150 plates (London, 1893); vol. ii, with atlas of 123 plates (London, 1898).

21. KENYON (F. G.), *Catalogue of Additions to the Department of Manuscripts in the British Museum*, 1888–1893, pp. 390–450 (descriptions of papyri cxxi–ccccLviii in the British Museum).

22. GRENFELL (B. P.), *Greek Papyri from Apollonopolis* (*Journal of Philology*, 1894).

23. GRENFELL (B. P.) and MAHAFFY (J. P.), *Revenue Laws of Ptolemy Philadelphus*, with thirteen plates (Oxford, 1896).

24. GRENFELL (B. P.), *An Alexandrian Erotic Fragment and other Greek Papyri* (Greek Papyri I), with one plate (Oxford, 1896).

25. GRENFELL (B. P.) and HUNT (A. S), *New Classical Fragments and other Greek and Latin Papyri* (Greek Papyri II), with five plates (Oxford, 1897).

26. GRENFELL (B. P.) and HUNT (A. S.), *The Oxyrhynchus Papyri*, Part I, with eight plates (Egypt Exploration Fund, London, 1898).

IV. GERMANY.

27. BOECKH (A.), *Erklärung einer ägyptischen Urkunde auf Papyrus* (*Abhandlungen der Berl. Akademie*, Berlin, 1821).

28. LETRONNE (J. A.), *Catalogue des antiquités découvertes en Egypte par M. Passalacqua*, with a facsimile (Paris, 1826).

29. SCHMIDT (W. A.), *Die griechischen Papyrusurkunden der königlichen Bibliothek zu Berlin*, with two facsimiles (Berlin, 1842).

30. PARTHEY (G. F. C.), *Zauberpapyri* (*Abhandlungen der königl. Akademie zu Berlin*, 1865).

31. PARTHEY (G. F. C.), *Die griechische Papyrusfragmente der Leipziger Universitätsbibliothek* (*Sitzungsberichte der königl. Akademie zu Berlin*, 1865).

32. WESSELY (K.), *Die griechischen Papyri Sachsens* (*Berichte der königl. Sächs. Gesellschaft der Wissenschaften*, 1885).

33. MAGIRUS (K.) and WESSELY (K.), *Griechische Papyri im ägyptischen Museum in Berlin* (in *Wiener Studien*, Vienna, 1886).

34. WILCKEN (U.). *Arsinoitische Steuerprofessionen und verwandte Urkunden*, with four plates (*Sitzungsberichte der königl. Akademie zu Berlin*, 1883).

35. WILCKEN (U.). *Actenstücke aus der königlichen Bank zu Theben in den Museen von Berlin, London, Paris* (*Abhandlungen der königl. Akademie zu Berlin*, 1886).

36. WILCKEN (U.), articles in *Hermes*, xix, xx, xxi. xxii, xxiii, xxvii, xxviii, xxix, xxx (1884–1895), and *Philologus*, liii (1894).

37. WILCKEN (U.), *Tafeln zur älteren griechischen Paläographie* (Leipzig and Berlin, 1891); twenty plates.

38. WILCKEN (U.), KREBS (F.), and VIERECK (P.), *Griechische Urkunden aus den Museen in Berlin*, vol. i, with three facsimiles (1892–1895), vol. ii, with facs. of Latin papyrus (1895–1898), vol. iii in progress.

V. HOLLAND.

39. REUVENS (C. J. C.), *Lettres à M. Letronne sur les Papyrus bilingues et grecs . . . du Musée d'Antiquités de l'Université de Leide* (Leyden, 1830).

40. LEEMANS (C.), *Papyri Graeci Musei Antiquarii Publici Lugduni-Batavi*, vol. i, with six plates (1843); vol. ii, with four plates (1885).

VI. ITALY.

41. SCHOW (N.), *Charta papyracea graece scripta Musei Borgiani Velitris*, with six plates (Rome, 1788).

42. FURIA (F. del), *Illustrazione di un papiro Greco, che si conserve presso il ch. sig. Luigi Lambruschini* (in Furia's *Collezione*, vol. 17, Florence, 1813).

43. PEYRON (A.), *Papyri Graeci Regii Taurinensis Musei Aegyptii*, with six plates (Turin, 1826).

44. MAI (A.), *Auctores Classici*, vol. iv. 445, v. 352, 356, 601, 602, 603 (Rome, 1831, 1832); texts of six papyri in the Vatican.

45. PEYRON (B.), *Papiri Greci . . . . della Bibliotheca Vaticana* (Turin, 1841) ; = no. 15 above. The Vatican papyri are four of those given by Mai, with commentary.

VII. SWITZERLAND.

46. NICOLE (J.), articles in *Revue de Philologie*, xvii, xx (1893, 1896), *Revue Archéologique*, xxiv, xxv (1893, 1894), *Revue des Etudes Grecques*, 1895.
47. NICOLE (J.), *Les Papyrus de Genève*, vol. i. fasc. 1 (1896).

APPENDIX IV

ABBREVIATIONS AND SYMBOLS

In the following list, A = the Brit. Mus. papyrus of Aristotle's Ἀθηναίων Πολιτεία.

H = Herculaneum Papyrus 157–152.

M = Medical Papyrus in the British Museum.

Tables of abbreviations and symbols used in non-literary papyri will be found in the indices to the *British Museum Catalogue of Papyri*, Grenfell's *Greek Papyri*, vols. i and ii, and the *Griechische Urkunden* of the Berlin Museum. All except those which are the caprices of an individual, or as to which some doubt attaches, are included in the list given below. In late theological papyri the familiar abbreviations θ̄ς. κ̄ς. ū̄ν̄ο̄ς κ.τ.λ. are found, as in contemporary vellum MSS.

SYMBOLS AND ABBREVIATIONS USED IN PAPYRI.

(a) IN LITERARY PAPYRI.

ϲ = αι (as part of a word) (A).

Ҩ = αἴτιος, αἰτία, κ.τ.λ. (M).

ἀˋ = ἀνά (A).

ϟ = αὐτός and cases (A and non-lit.).

γ′ = γάρ (A, M).

γˋ = γάρ (H).

Г̔ = γίνεται κ.τ.λ. (M).

δ′ = δέ (A).

δˋ = διά (A).

δ^υ = δύναμις and cases (M).

\ = εἶναι (A, H, M).

⟩ = εἰσίν (A, M).

/ = ἐστί (A, H, M).

θ′ = θαι (A).

κ′ = και (A).

κˋ = και (H).

κ̂ = κατά (M).

κˋ = κατά (A).

λ̣ = λόγος (M, magical papp.).

μ′ = μεν (A, M).

μˋ = μετά (A).

N̄ = νον (M).

N̂ = νης (M).

Ṅ = νων (M).

o' = ουν (A).

ō = οὔτω (M).

π̀ = παρά (A, M).

π' = περ or περί (A).

π' = πρός (M).

▯ = ποιητής (Harris Homer).

ᵐ̄ = πρότερον (M).

ᵐ̄ = πρός (H, Brit. Mus. Pap. CCLVI *verso*).

σ' = σύν (A).

τ̀ = την (A).

τ' = της (A).

ϝ = τρόπος and cases (H).

τ' = των (A, M).

τ̀ = των (H).

ῦ = ὑπάρχει κ.τ.λ. (M).

υ' = ὑπέρ (A).

ῠ = ὑπό (A).

φ = φησίν, κ.τ.λ. (M).

✱ = χρόνος (A, H).

(b) IN NON-LITERARY PAPYRI.

ⅆ = ἀπό.

≈ = ἄρουρα (rare).

ɤ = ἄρουρα (common).

⟨ = ἀρτάβη (rare).

ϛ = ἀρτάβη (rare).

—₀ = ἀρτάβη (common).

5 = αὐτός and cases.

Ⴝ = δεῖνα (in magical papp.).

Ⱶ = δραχμή (Ptolemaic).

⟨ = δραχμή (Roman).

ʃ = δραχμή (Roman, less common).

ʒ = δραχμή (Roman).

L = ἔτος and cases.

δ = ἥλιος and cases (magical papp.).

Ꝫ = καί (?).

s = καί.

4 = κεράτιον.

Ꞃ = μετρητής.

▯, ▢, or ▯μα = ὄνομα (magical papp.).

▢, ▢▢, ▢▢, or ▯τα = ὀνόματα (magical papp.).

∩ = πῆχυς.

▯ = ποίημα (magical papp.).

▯ = πρᾶγμα (magical papp.).

∩ = πόλις.

Ⱶ = πυροῦ, or loosely πυροῦ ἀρτάβη.

℄ = σελήνη and cases (magical papp.).

§ = σκῆπτρον (magical papp.).

Ā = τάλαντον.

⟨ = ὑπέρ (Byzantine).

χ^L = χαλκοῦ.

✱ = χρῖε, κ.τ.λ. (magical papp.).

— = 1 obol.

= = 2 obols.

Γ = 3 obols.

F = 4 obols.

Ƒ = 5 obols.

χ^o = 2 chalchi.

o' = 4 chalchi or ½ obol.

$o'\chi^o$ = 6 chalchi.

ϛ = 90.

9 = 90.

T = 900.

ℋ = $\frac{3}{4}$ [1].

o' = $\frac{2}{3}$.

ᒪ = $\frac{1}{2}$.

ᑐ = $\frac{1}{4}$ [2].

⨼ = $\frac{1}{8}$.

/ = γίνεται, or total.

Γ = γίνεται, or total.

L = sign of subtraction.

Ɔ = remainder.

∩ = remainder (περίεστι).

---

[1] A combination of the signs for $\frac{1}{2}$ and $\frac{1}{4}$.

[2] The explanation of this sign is given by Brit. Mus. Pap. ɒcɪv; it is the letter Δ (= 4) written like o (as often in numerals, e. g. έ́ό constantly = ό̣,), and with a stroke above it to mark that it is a fraction (ό, hence d). A δ of the form d does not appear till the fourth century.

INDEX

——◆——

THE END

www.ingramcontent.com/pod-product-compliance
Lightning Source LLC
Chambersburg PA
CBHW021706210326
41599CB00013B/1543